Methods in Molecular Biology

Series Editor
John M. Walker
School of Life and Medical Sciences
University of Hertfordshire
Hatfield, Hertfordshire, AL10 9AB, UK

For further volumes:
http://www.springer.com/series/7651

Oat

Methods and Protocols

Edited by

Sebastian Gasparis

Department of Functional Genomics, Plant Breeding & Acclimatization Inst.
National Research Institute, Blonie, Poland

Humana Press

Editor
Sebastian Gasparis
Department of Functional Genomics
Plant Breeding & Acclimatization Inst.
National Research Institute
Blonie, Poland

ISSN 1064-3745 ISSN 1940-6029 (electronic)
Methods in Molecular Biology
ISBN 978-1-4939-6680-6 ISBN 978-1-4939-6682-0 (eBook)
DOI 10.1007/978-1-4939-6682-0

Library of Congress Control Number: 2016961427

Photo Credit: Yuliya Yanushevska

Printed on acid-free paper

This Humana Press imprint is published by Springer Nature
The registered company is Springer Science+Business Media LLC
The registered company address is: 233 Spring Street, New York, NY 10013, U.S.A.

Preface

Oat is a cereal grain belonging to the genus *Avenae* which contains more than 20 species occurring at three ploidy levels, diploid, tetraploid, and hexaploid. The primary cultivated oat (*Avena sativa* L.) is a hexaploid ($2n=6x=42$) consisting of three genomes, AA, CC, and DD. Hexaploid oat is cultivated worldwide in all agricultural regions which have a moderate climate and is used mainly for production of feed and food. With respect to the production quantity, from an economic standpoint oat is significantly less important when compared to main cereals such as rice, maize, wheat, and barley; however, the particular chemical composition of oat grains makes it unique among other cereal species. The most important properties of oat grain are a high level of proteins rich in exogenous amino acids, high levels of unsaturated fatty acids, and finally the presence of beta-glucan, which has been shown to have a positive effect on the prevention of diseases of modern civilization. The favorable nutritional properties of oat make it an interesting object for breeders and scientists, especially with respect to genetic and genomic studies, where the progress compared to other cereals is still not satisfactory. The insufficient progress in these genomic studies is caused not only by its lower economic importance, but the size and complexity of the hexaploid oat genome is a serious constrain for the application of conventional research tools.

The recent advances in molecular biology and the fusion of various genetic engineering techniques have enabled many scientific breakthroughs in plant science and led to the development of new "reverse genetics" approaches, such as RNA interference technology, and new genome editing techniques. The lack of detailed, robust protocols is one of the reasons that the modern research approaches have not yet been applied in the genomic studies of oat.

This book provides detailed protocols which have been developed or modified exclusively for oat by the internationally recognized experts. Each chapter describes, step-by-step, the application of state-of-the-art technology which may be useful for genetic studies of oat. The protocols are a selection of various molecular biology and biotechnology methods including the application of molecular markers for polymorphism analyses, cytological manipulations and the production of synthetic polyploids, and in vitro cultures and genetic modification. The authors hope that this book will contribute to the development of new research approaches that reveal the organization of the oat genome and allow the identification of new useful traits for further improvement of this exceptional crop.

I would like to thank all the contributing authors for sharing their knowledge and experience. Special thanks go to Professor John Walker for his guidance and support.

Błonie, Poland ***Sebastian Gasparis***

Contents

Part IV Genomic Studies

Part V Proteomics Methods

Contributors

DENIZ AKDEMIR • *Plant Breeding and Genetics Section, School of Integrative Plant Science, Cornell University, Ithaca, NY, USA*
MAJA BOCZKOWSKA • *National Centre for Plant Genetic Resources, Plant Breeding and Acclimatization Institute – National Research Institute, Błonie, Poland*
COREY D. BROECKLING • *Proteomics and Metabolomics Facility, Colorado State University, Fort Collins, CO, USA*
PHILIP A. DAVIES • *South Australian Research and Development Institute, Adelaide, SA, Australia*
RICHARD DIETRICH • *Chair of Hygiene and Technology of Milk, Ludwig-Maximilians University Munich, Oberschleissheim, Germany*
MARTA DZIUBA • *Chair of Food Biochemistry, University of Warmia and Mazury in Olsztyn, Olsztyn, Poland*
YONG-BI FU • *Plant Gene Resources of Canada, Saskatoon Research and Development Centre, Agriculture and Agri-Food, Saskatoon, SK, Canada*
DAVID F. GARVIN • *USDA-ARS Plant Science Research Unit, St. Paul, MN, USA*
SEBASTIAN GASPARIS • *Department of Functional Genomics, Plant Breeding and Acclimatization Institute—National Research Institute, Błonie, Poland*
JUAN J. GUTIERREZ-GONZALEZ • *Department of Agronomy and Plant Genetics, University of Minnesota, St. Paul, MN, USA*
OWEN A. HOEKENGA • *Plant Genomics Consultant, Ithaca, NY, USA*
TAKAYOSHI ISHII • *Leibniz Institute of Plant Genetics and Crop Plant Research (IPK), Stadt Seeland, Germany*
JULIO ISIDRO-SÁNCHEZ • *Agriculture and Food Science, University College Dublin, Dublin, Ireland*
ELINA KIVIHARJU • *Natural Resources Institute Finland, Green Technology, Jokioinen, Finland*
DIETMAR KNOPP • *Institute of Hydrochemistry and Chair of Analytical Chemistry, Technical University Munich, Munich, Germany*
VERONICA M.T. LATTANZIO • *Institute of Sciences of Food Production (ISPA), National Research Council of Italy (CNR), Bari, Italy*
ERWIN MÄRTLBAUER • *Chair of Hygiene and Technology of Milk, Ludwig-Maximilians University Munich, Oberschleissheim, Germany*
NICOLE MIHELICH • *University of Wisconsin-Madison, Madison, WI, USA*
PIOTR MINKIEWICZ • *Chair of Food Biochemistry, University of Warmia and Mazury in Olsztyn, Olsztyn, Poland*
SIRPA MOISANDER • *Natural Resources Institute Finland, Green Technology, Jokioinen, Finland*
GRACIA MONTILLA-BASCÓN • *Plant Breeding and Genetics Section, School of Integrative Plant Science, Cornell University, Ithaca, NY, USA*
TOSHINOBU MORIKAWA • *Graduate School of Life and Environmental Sciences, Osaka Prefecture University, Osaka, Japan*

DOROTA NAŁĘCZ • *Chair of Food Biochemistry, University of Warmia and Mazury in Olsztyn, Olsztyn, Poland*
REINHARD NIESSNER • *Institute of Hydrochemistry and Chair of Analytical Chemistry, Technical University Munich, Munich, Germany*
NOAN NIVARLET • *UNISENSOR S.A., Zoning industriel du Dossay, Liège, Belgium*
AGNIESZKA ONYŚK • *Department of Functional Genomics, Plant Breeding and Acclimatization Institute – National Research Institute, Błonie, Poland*
SUSANNA OSWALD • *Institute of Hydrochemistry and Chair of Analytical Chemistry, Technical University Munich, Munich, Germany*
MICHELANGELO PASCALE • *Institute of Sciences of Food Production (ISPA), National Research Council of Italy (CNR), Bari, Italy*
ELENA PRATS • *Institute for Sustainable Agriculture, CSIC, Córdoba, Spain*
AURÉLIE M. RAKOTONDRAFARA • *University of Wisconsin-Madison, Madison, WI, USA*
ROBYN ROBERTS • *University of Wisconsin-Madison, Madison, WI, USA*
DANIELLE SAVINO • *University of Wisconsin-Madison, Madison, WI, USA*
PARMINDER K. SIDHU • *South Australian Research and Development Institute, Adelaide, SA, Australia*
MARK SORRELLS • *Plant Breeding and Genetics Section, School of Integrative Plant Science, Cornell University, Ithaca, NY, USA*
IWONA SZERSZUNOWICZ • *Chair of Food Biochemistry, University of Warmia and Mazury in Olsztyn, Olsztyn, Poland*
PIRJO TANHUANPÄÄ • *Natural Resources Institute Finland, Green Technology, Jokioinen, Finland*
EVA WEGEL • *Cell and Developmental Biology, John Innes Centre, Norwich, UK*
MO-HUA YANG • *Plant Gene Resources of Canada, Saskatoon Research and Development Centre, Agriculture and Agri-Food, Saskatoon, SK, Canada; College of Forestry, Central South University of Forestry and Technology, Changsha, Hunan, China*
JINCAN ZHANG • *University of Wisconsin-Madison, Madison, WI, USA*

Part I

Cytological Methods

Chapter 1

Fluorescence In Situ Hybridization in Oat

Eva Wegel

Abstract

This chapter describes methods to detect gene loci or gene transcripts by fluorescence labeling. Fluorescence in situ hybridization (FISH) can be used to identify the positions of genes or BACs or the distribution of repetitive sequences on metaphase chromosomes as well as the identification of alien chromosomes. It enables the identification of gene loci and active transcription sites in interphase nuclei and also the localization of cellular transcripts. The protocols here deal with the production of DNA and RNA probes, the preparation of oat metaphase spreads and root tissue sections, the subsequent hybridization, post-hybridization washes, and detection by immunofluorescence.

Key words FISH, Nick translation, In vitro transcription, Metaphase spreads, Wax embedding, Immunofluorescence

1 Introduction

In fluorescence in situ hybridization (FISH) a probe that contains a modified nucleotide in several positions binds to a complimentary nucleotide sequence. The modified nucleotide is then detected by immunofluorescence. FISH can be used (a) to detect the chromosomal locations of genes, BACs, cosmids, and repetitive sequences or to identify whole chromosomes (DNA FISH) and (b) to detect transcripts both at transcription sites in the nucleus and in the cytoplasm (RNA FISH). Classically DNA FISH has been employed to identify which chromosome carries a particular gene or BAC sequence in a metaphase spread, to identify chromosomal deletions or translocations, or to identify chromosomes from the two parents in hybrids (Fig. 1a). In interphase nuclei, DNA FISH has been applied to determine the positions of individual chromosomes or gene loci within the nuclear volume or to measure transcription related chromatin decondensation (Fig. 1b, c) [1, 2]. RNA FISH is used to identify cells and cell types that express a certain transcript or to identify at which developmental stage a

Sebastian Gasparis (ed.), *Oat: Methods and Protocols*, Methods in Molecular Biology, vol. 1536,
DOI 10.1007/978-1-4939-6682-0_1, © Springer Science+Business Media LLC 2017

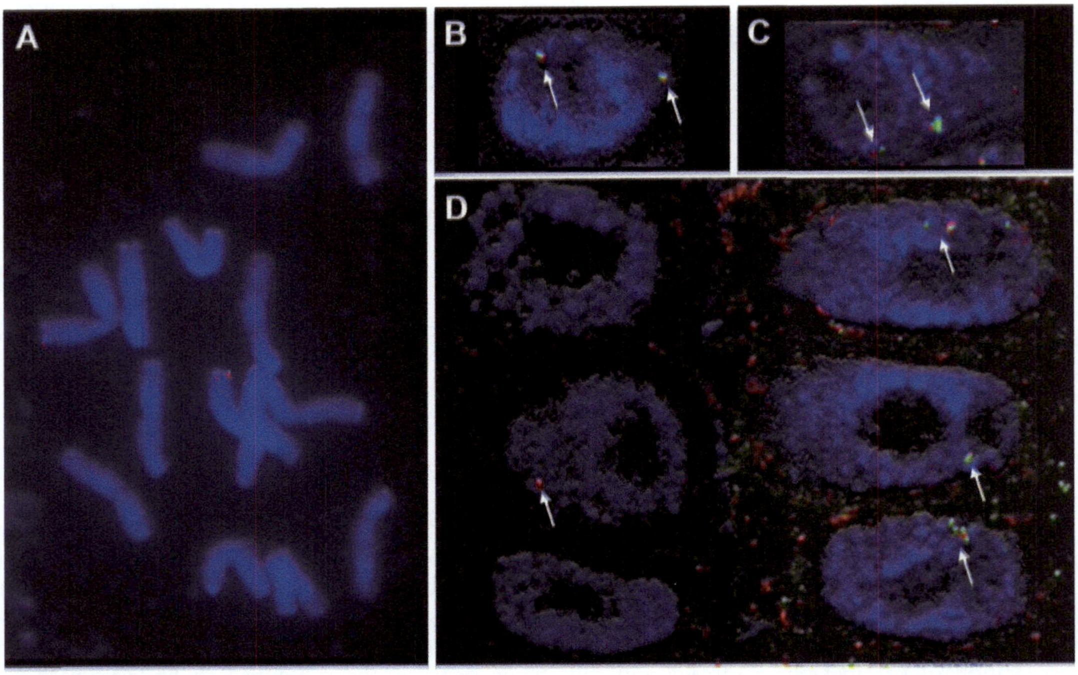

Fig. 1 Examples of DNA and RNA FISH in oat. (**a**) Metaphase plate showing the position of the Sad1 gene (*red*). (**b** and **c**) DNA FISH on G1 nuclei (*blue*) in oat roots reveals the positions of the Sad1 (*red*) and Sad2 (*green*) gene (*arrows*). Both genes are highly expressed in the epidermis (**c**), while Sad2 is not expressed in the subepidermis at all and Sad1 to a lesser degree (**b**). As a result the 80 kb region spanning the two genes is decondensed in the epidermis and condensed in the subepidermis. (**d**) RNA FISH reveals nascent transcripts of Sad1 (*red*) and Sad2 (*green*) in nuclei of the subepidermis on the left and the epidermis on the right (*arrows*). The cytoplasmic localizations of the two transcripts in epidermis cells do not overlap and there is no Sad2 transcript in the subepidermis. (**b–d**) Overlay of several optical sections

gene is expressed and even to see how many alleles are actively transcribed within a nucleus (Fig. 1d) [1, 2]. Using fluorescence detection the cytoplasmic localization of individual transcripts can be resolved at subcellular level [2]. This is in contrast to detection by color precipitation, which cannot differentiate between different subcellular localizations. It is possible to combine DNA and RNA FISH by using flanking sequences to label the gene and coding sequences to label the transcript [1, 2].

DNA probes are used for DNA FISH and RNA probes for RNA FISH (Subheadings 3.1 and 3.2). During hybridization denatured DNA probes are supposed to bind to both strands of the denatured target genomic sequences. In RNA FISH, the probe (an antisense transcript) forms a double strand with the sense target transcript in the cell. In principle, RNA probes could also be used for DNA FISH as the stability of double-stranded nucleotide sequences increases from DNA–DNA to DNA–RNA to RNA–RNA. However, single-stranded RNA probes are susceptible to degradation by RNAses, which are ubiquitous, so care must be taken to work in as clean an environment as possible and DNA in situ hybridization experiments commonly include an RNase step to remove transcripts, which would increase background hybridization. DNA probes are usually prepared by nick translation. Nick translation mixes or kits are commercially available and their use is recommended. RNA probes are prepared by in vitro transcription, for which enzyme mixes are available and their yield is significantly greater than that of nick translation reactions. Genomic sequences down to 5 kb can be detected by DNA FISH and transcript probes should be at least 500 bp in length.

Subheading 3.4 describes how to prepare metaphase spreads from oat roots. Diploid *Avena strigosa* was used for this experiment, but it will also work with roots from hexaploid oat seedlings. The metaphase spreads are digested with RNase to remove any transcripts that could cause background hybridization and with pectinase to remove histones and other proteins that could hinder hybridization to the highly condensed chromatin. After a post-fixation step both probe and chromatin are denatured and hybridized overnight (o/n, Subheading 3.5). Subheading 3.9 describes the post-hybridization stringency washes to remove unspecifically bound probe and Subheading 3.11 describes the immunodetection.

Tissue sections were also prepared from root tips of *Avena strigosa* seedlings. Other tissues might need modifications to the protocol where indicated. For both DNA and RNA FISH, tissue is harvested and fixed in freshly prepared 4% formaldehyde, which is superior to commercial formaldehyde solutions. It is then embedded in wax in a tissue processing machine and sectioned on a rotary microtome (Subheading 3.6).

For DNA FISH, tissue sections are dewaxed and treated with cell wall degrading enzymes to facilitate probe and antibody penetration (Subheading 3.7). An RNAse step removes any transcripts that could cause background hybridization. After that both probe and chromatin are denatured and hybridized o/n. Subheading 3.9 describes the post-hybridization stringency washes to remove nonspecifically bound probe and Subheading 3.11 describes the immunodetection.

For RNA FISH, tissue sections are also dewaxed and treated with cell wall degrading enzymes to facilitate probe and antibody penetration (Subheading 3.8). Proteinase K treatment removes any proteins that coat the transcripts and hinder probe access. After a post-fixation step the probe is denatured to remove possible hairpin structures and the slides are hybridized o/n. Subheading 3.10 describes the post-hybridization stringency washes for RNA and Subheading 3.11 describes the immunodetection.

For imaging of metaphase spreads and semithin tissue sections, a widefield microscope with a good-quality cooled CCD or an sCMOS camera is preferable to a scanning confocal microscope because the sensitivity of the former exceeds that of even modern GaAsP or hybrid detectors, which is important for weak signals from small or low-abundancy probes. When using a widefield microscope, deconvolution, which removes out of focus blur, will greatly improve the quality of image stacks from tissue sections and should always be used. If deconvolution software is not available, then a confocal microscope might give better results. Metaphase spreads do not need deconvolution.

2 Materials

2.1 Preparation of DNA Probes

1. Heating/cooling block or 15 °C water bath.
2. Nick translation mix (Roche).
3. Digoxigenin-11-dUTP (Roche) or dinitrophenol-11-dUTP (PerkinElmer) or biotin-16-dUTP (Roche).
4. dNTP-Set, 100 mM each of the four dNTPs.
5. 1 kb DNA ladder.
6. 3 M Na acetate, pH 5.2.

2.2 Preparation of RNA Probes

1. PCR machine.
2. T7 RNA polymerase (Roche, comes with transcription buffer).
3. Sterile deionized water.
4. NTP mix: Make up in sterile deionized water using the following final concentrations: 10 mM ATP, 10 mM CTP, 10 mM GTP, 6.5 mM UTP (any supplier) and 3.5 mM of either biotin-16-UTP (Roche) or digoxigenin-11-UTP (Roche) or dinitrophenol-11-UTP (PerkinElmer).
5. RNAsin ribonuclease inhibitor (Promega).
6. DNase I (RNase free), 10 U/µL (Roche, *see* **Note 1**).
7. 200 mM EDTA, pH 8.0, autoclaved.
8. 4 M LiCl, autoclaved.
9. Filter-sterilized 200 mM carbonate buffer, pH 10.2: 80 mM $NaHCO_3$ and 120 mM Na_2CO_3, no need to pH. Freeze in aliquots for single use.
10. 10% acetic acid.
11. 3 M Na Acetate, pH 5.5, autoclaved.
12. 1 kb DNA ladder.

2.3 Preparation of 4% Formaldehyde

1. pH 4.5–10 indicator strips.
2. Paraformaldehyde prills (Sigma-Aldrich, carcinogen!) (*see* **Note 2**).

3. Dilute H_2SO_4: 10% (v/v) solution of H_2SO_4. Prepare by drop-wise addition of concentrated (98%) sulfuric acid (toxic, causes severe burns!) to deionized water under the fume hood.

2.4 Preparation of Metaphase Spreads

1. Stereo microscope.
2. Pastettes (plastic Pasteur pipettes).
3. Straight no. 5 forceps (e.g., from TAAB) or dissection needles.
4. Coverslips No. 1, 18×18 mm.
5. Microscope equipped with a 10× or 20× objective and phase contrast or DIC optics.
6. Single edge razor blades.
7. Fixative: 3 parts ethanol plus 1 part acetic acid.
8. 100 mL 10× FISH buffer: 60 mL of 0.1 M tri-sodium citrate+40 mL of 0.1 M Citric Acid. Store at 4 °C.
9. Cell wall digestion mixture for metaphases: 2% (w/v) 'Onozuka' R10 cellulase from *Trichoderma viride* (Yakult, Japan), 20% (v/v) pectinase from *Aspergillus niger* (Sigma P4716) in FISH buffer. Prepare 2 mL of the digestion mixture by adding 40 mg of 'Onozuka' R10 cellulase and 400 μL pectinase to 1.6 mL 1× FISH buffer. Weigh out in the fume hood as these enzymes are potential sensitizers. Store in 50 μL aliquots at −20 °C.
10. 45% acetic acid: glacial acetic acid diluted in deionized water to a concentration of 45%.
11. Dry ice.

2.5 Pretreatment of Metaphase Spreads and Hybridization for DNA FISH

1. Slide denaturation/hybridization system (Senova) or in situ slide adapter for a PCR machine.
2. RNAse A (Sigma): 10 mg/mL stock solution in deionized water, filter-sterilized and frozen at −20 °C in aliquots, which can be defrosted several times. The working solution of 10 μg/mL in 2× SSC is prepared fresh.
3. Pepsin from porcine gastric mucosa (Sigma, lyophilized powder, 4720 U/mg protein): Stock solution 500 μg/mL in 0.01 M hydrochloric acid, dilute 1:20 in 0.01 M HCl for the working solution, which can be kept in single use aliquots at −20 °C.
4. 4% Formaldehyde: *see* Subheadings 2.3 and 3.3.
5. Formamide, deionized, high grade (Toxic! *see* **Note 3**). Store in small aliquots at −20 °C.
6. 50% (w/v) dextran sulfate, sodium salt, molecular biology grade. Prepare the solution by mixing dextran sulfate with deionized water and heating it up to 65 °C until completely dissolved. Freeze immediately in small aliquots and store at −20 °C.

7. 10% SDS.
8. Salmon sperm DNA (Sigma), 5 μg/μL, autoclaved.

2.6 Wax Embedding of Oat Roots for DNA and RNA FISH

1. Vacuum desiccator with pump.
2. Tissue Tek VIP tissue processor (Sakura).
3. Tissue or biopsy cassettes.
4. Paraffin wax embedding station.
5. Rotary microtome (Leica).
6. Poly-L-lysine coated microscope slides.
7. Slide warmer.
8. 4% formaldehyde: *see* Subheadings 2.3 and 3.3.
9. Ethanol for tissue processor.
10. Xylene (general purpose grade) for tissue processor.
11. Paramat wax for tissue processor.

2.7 Pretreatment and Hybridization of Root Sections for DNA FISH

1. Slide denaturation/hybridization system (Senova) or in situ slide adapter for a PCR machine.
2. Histolene (Cellpath).
3. Enzyme mix: 1% (w/v) driselase (Sigma, contains laminarinase, xylanase, cellulase activity), 0.5% (w/v) cellulase "Onozuka" R10 (Yakult, Japan), 0.025% (w/v) pectolyase Y23 (Sigma) in PBS. Weigh out in the fume hood as these enzymes are potential sensitizers. Aliquots of the mix can be kept frozen at −20 °C (*see* **Note 4**).
4. RNase A (Sigma, *see* Subheading 2.5).
5. Formamide, deionized, high grade (Toxic! *see* **Note 3**). Store in small aliquots at −20 °C.
6. 50% dextran sulfate, molecular biology grade (*see* Subheading 2.5).
7. 10% SDS.
8. Salmon sperm DNA (Sigma), 5 μg/μL, autoclaved.

2.8 Pretreatment and Hybridization of Root Sections for RNA FISH

1. Oven at 30–50 °C.
2. Glass Coplin jars, autoclaved.
3. Histolene (Cellpath).
4. Enzyme mix: 1% driselase (Sigma-Aldrich, D9515), 0.5% cellulase "Onozuka" R10 (Yakult, Japan), 0.025% pectolyase Y23 (Sigma) in PBS. Weigh out in the fume hood as these enzymes are potential sensitizers. Aliquots of the mix can be kept frozen at −20 °C (*see* **Note 4**).
5. Proteinase K buffer: 100 mM Tris–HCl pH 8.0, 50 mM EDTA, autoclaved.

6. Proteinase K, molecular biology grade (Sigma): 10 mg/mL stock in deionized water kept in aliquots at −20 °C.
7. 50 mg/mL glycine stock in deionized water, filter-sterilized and kept at 4 °C.
8. 4% formaldehyde (*see* Subheadings 2.3 and 3.3).
9. Formamide, deionized, high grade (Toxic! *see* **Note 3**). Store in aliquots at −20 °C.
10. tRNA, 50 μg/μL (Sigma type XXI, made up in deionized water).
11. 10× salts (amounts for 1 L): 3 M NaCl (175.3 g), 0.1 M Tris–HCl pH 6.8 (100 mL of a 1 M sol.), 7.8 g $NaH_2PO_4 \cdot 2H_2O$ + 7.1 g Na_2HPO_4 (=0.1 M $NaPO_4$ buffer), 50 mM EDTA (100 mL of a 0.5 M sol.).
12. Hybridization buffer (800 μL, stored at −20 °C) (*see* **Note 3**):

    100 μL 10× salts.

    400 μL formamide, deionized.

    200 μL 50% dextran sulfate, molecular biology grade (*see* Subheading 2.5, *see* **Note 5**).

    20 μL 50× Denhardt's solution (Sigma).

    80 μL H_2O.

2.9 Post-hybridization Washes for DNA FISH

1. 20% formamide in 0.1× SSC: 20 mL formamide (general purpose reagent (Toxic! *see* **Note 3**)) + 80 mL deionized water + 0.5 mL 20× SSC.
2. 4× SSC, 0.2% (v/v) Tween 20.

2.10 Post-hybridization Washes for RNA FISH

1. 2× SSC/50% formamide: 50 mL formamide (general purpose reagent (Toxic! *see* **Note 3**)) + 40 mL deionized water + 10 mL 20× SSC.
2. 1× SSC/50% formamide: 50 mL formamide (general purpose reagent) + 45 mL deionized water + 5 mL 20× SSC.
3. 2× SSC.
4. 4× SSC, 0.2% (v/v) Tween 20.

2.11 Immunodetection

1. Blocking solution: 5% (w/v) bovine serum albumin (BSA, fraction V), 4× SSC, 0.2% Tween 20.
2. Primary antibodies: mouse monoclonal anti digoxin antibody (Sigma), rabbit anti-dinitrophenyl antibody (Invitrogen).
3. ExtrAvidin-Cy3 (Sigma).
4. Secondary antibodies: Alexa Fluor 488 goat anti mouse antibody (Invitrogen), anti rabbit Alexa Fluor 647 antibody (Invitrogen).
5. 4′,6-Diamidino-2-phenylindole (DAPI, Sigma): stock solution 1 mg/mL in deionized water kept at −20 °C (*see* **Note 6**).

6. Vectashield antifade mounting medium (Vector Laboratories H-1000) or Prolong Diamond (ThermoFisher).
7. Nail polish.

2.12 Repeatedly Used Equipment and Reagents

1. Heat block with stirring function.
2. Bench top centrifuge.
3. Gel electrophoresis equipment.
4. Plastic coverslips made from autoclave bags: transparent autoclave bags are cut into squares the size of coverslips (*see* **Note** 7).
5. Humid chamber: Line the bottom of a lidded plastic box with wet tissue paper. Place the slides on toothpicks or similar to lift them off the paper.
6. Coplin jars.
7. Incubator at 37 °C.
8. Shaker.
9. Fume hood.
10. Water bath, for some steps a shaking function is also required.
11. 30%, 50%, 70% and absolute ethanol.
12. TE: 10 mM Tris–HCl pH 8.0, 1 mM EDTA, autoclaved.
13. 10× PBS, pH 7.4: 1.3 M NaCl, 0.07 M Na_2HPO_4, 0.03 M NaH_2PO_4, adjust pH with 5 M NaOH or 5 M HCl, autoclave.
14. 20× SSC, pH 7.0: 3 M NaCl, 300 mM tri-sodium citrate, adjust pH with 1 M HCl, autoclave.

3 Methods

3.1 Preparation of DNA Probes

1. Use the manufacturer's instructions that come with the nick translation mix to prepare DNA probes (*see* **Note 8**).
2. Check the size of the probes by loading a 2 μL aliquot next to a 1 kb DNA ladder on a 1% agarose gel. The probe should produce a smear between 200 and 500 bp in length.
3. Add 1/10 volume of 3 M Na Acetate pH 5.2 and 2.5 volumes of ice-cold ethanol and precipitate at −20 °C for at least 1 h.
4. Centrifuge for 15 min at 4 °C at 13,200 rpm.
5. Wash with ice-cold 70% ethanol and spin again, remove supernatant and air-dry for 5 min.
6. Resuspend pellet in 10–20 μL TE. Probes keep for months at −20 °C.

3.2 Preparation of RNA Probes

1. Synthesize DNA templates by PCR with a modified reverse primer that has the T7 promoter added to its 5′ end. This way

the T7 polymerase transcribes the antisense transcript. Precipitate linear PCR products and resuspend them in sterile deionized water to a concentration of 0.5–1 μg/μL (*see* **Note 9**).

2. Set up a 20 μL in vitro transcription reaction by pipetting the reagents together at room temperature (!) (*see* **Note 10**):

   11 μL sterile deionized water.

   2 μL 10× transcription buffer (thaw at room temperature).

   2 μL NTP mix.

   1 μL RNAsin ribonuclease inhibitor (20 U).

   2 μL DNA template (0.5–1 μg).

   2 μL T7 RNA polymerase (40 U).

3. Incubate at 37 °C for 2 h.
4. Add 2 μL DNase I (RNase free), mix well, spin and incubate at 37 °C for 15 min.
5. Add:

   2 μL 200 mM EDTA, pH 8.0.

   2 μL LiCl (4 M).

   75 μL ice-cold ethanol.

6. Mix and precipitate at −20 °C o/n.
7. Centrifuge for 15 min at 4 °C at 20,000 × *g*.
8. Wash with ice-cold 70% ethanol and spin again, remove supernatant and air-dry for 5 min.
9. Resuspend pellet in 50 μL sterile deionized water, add 50 μL of 200 mM carbonate buffer and mix.
10. Incubate at 60 °C for required time, which depends on probe length according to the following equation:

    $t = (L_i - L_f)/K \times L_i \times L_f$

    where t = time in minutes, K = rate constant (=0.11 kb/min), L_i = initial length (kb) and L_f = final length (optimal L_f = 0.15 kb)

11. Stop the reaction: add 5 μL acetic acid (10%), 10 μL Na Acetate (3 M, pH 5.5), 288 μL ethanol (ice-cold) and precipitate for 2 h to o/n at −20 °C.
12. Spin and wash and air-dry pellet as above.
13. Resuspend pellet in 50 μL TE.
14. Check probe concentration by loading 3 μL alongside a 1 kb DNA ladder on a 1% agarose gel. You should see a smear in the region of 50–200 bp when loading the probe alongside a 1 kb DNA ladder. Probes keep for months at −20 °C.

3.3 Preparation of 4 % Formaldehyde

1. Carry out every step of the procedure in the fume hood as formaldehyde is a carcinogen. Weigh out 4 g of paraformaldehyde for a final volume of 100 mL and transfer to a beaker containing a stirrer bar. Add 50 mL of deionized water.
2. Heat mixture with constant stirring on a hot plate to 60 °C and as it is warming up add 250 μL of 1 M NaOH. Check the temperature of the beaker and do not exceed 60 °C as the formaldehyde will break down otherwise.
3. When the paraformaldehyde has completely dissolved, turn off the heat and add 40 mL deionized water and 10 mL 10× PBS.
4. Let the solution cool down to room temperature (on ice) and add dilute H_2SO_4 in μL steps while measuring the pH using pH strips. Stop when the pH has reached 7.4 (*see* **Note 11**).

3.4 Preparation of Metaphase Spreads

1. Germinate oat seeds on filter paper and allow the roots to grow to a length of 2–3 cm.
2. Cut off 1–2 cm long pieces containing the roots tips and place three each in microfuge tubes with ice water, i.e., water that contains ice crystals. Leave the microfuge tubes in an ice bucket in the cold room for 24 h.
3. Prepare fixative (3 parts ethanol plus 1 part acetic acid) and transfer three root tips each to microfuge tubes with 1 mL of fixative. Leave at room temperature o/n. Soak slides for sample spreading in absolute ethanol o/n (Samples can be stored in fixative at −20 °C).
4. Rinse roots in 1× FISH buffer 3× for 5 min in a small petri dish on the shaker to remove all traces of fixative.
5. Place three roots in a microfuge tube containing 50 μL of the cell wall digestion mixture and incubate at 37 °C for 45–90 min. When the tips are dangling down the roots are ready (*see* **Note 12**).
6. Fill the microfuge tube with 1× FISH buffer and use a Pastette to transfer the roots to a small petri dish.
7. Transfer a single root to an ethanol cleaned slide using a Pastette.
8. Remove excess liquid carefully with filter paper and pipette 30 μL of 45 % acetic acid on to the root.
9. Heat the drop to 40–50 °C on a heat block.
10. Under a stereo microscope remove the root cap, tease out the terminal 1–3 mm, i.e., the dividing zone, and discard the elongation zone above.
11. Gently macerate the dividing zone with no. 5 forceps or dissection needles.
12. Carefully cover with a coverslip (No. 1, 18 × 18 mm) excluding air bubbles, put filter paper on top and tap gently to disperse cells.

13. Check metaphases under a microscope (phase contrast 1 or DIC, 10–20× objective).
14. Repeat heating the slide until the cytoplasm has been removed and cells are cleared.
15. Repeat gently tapping the coverslip and checking under the microscope until metaphase chromosomes are well spread outside the cells.
16. Use your thumb to press down hard on the filter paper covered coverslip, transfer the slide to dry ice and leave for 10 s; as soon as the coverslips are white they can be removed with a razor blade.
17. Air-dry slide (*see* **Note 13**).
18. Proceed to Subheading 3.5.

3.5 Pretreatment of Metaphase Spreads and Hybridization for DNA FISH

1. Treat slides with an RNAse A working solution of 10 μg/mL in 2× SSC. Apply 100–200 μL per slide, cover the metaphase spreads with plastic coverslips and incubate the slides in a humid chamber for 1 h at 37 °C. During the incubation prepare 4% paraformaldehyde (Subheading 3.3).
2. Remove coverslips and wash slides 3× 5 min in 2× SSC in a Coplin jar without shaking.
3. Treat slides with pepsin working solution (1:20 dilution of the stock in 0.01 M hydrochloric acid). Apply 100–200 μL per slide and incubate for exactly 10 min in a humid chamber at 37 °C.
4. Wash slides for 5 min in 2× SSC in a Coplin jar without shaking.
5. Fix slides in 4% paraformaldehyde for 2× 5 min in a Coplin jar on a shaker in the fume hood at RT.
6. Wash 3× 5 min in 2× SSC in a Coplin jar on a shaker at RT.
7. Dehydrate slides on a shaker for 3 min in 70% ethanol followed by 3 min in absolute ethanol.
8. Air-dry slides for at least 1 h (*see* **Note 14**).
9. Prepare hybridization solution, 30 μL per slide for an 18 mm × 18 mm area, amounts given for two slides:

   30 μL formamide, deionized, high grade.

   12 μL 50% dextran sulfate (*see* **Note 5**).

   6 μL 20× SSC.

   1 μL 10% SDS.

   2 μL salmon sperm DNA.

   4 μL DNA probe, 40 ng/μL (nick translation Subheading 3.1 or deionized water as a no-probe control (*see* **Note 15**)).

   5 μL H_2O.

10. Denature for 5 min in a water bath at 95 °C and then cool on ice for 5 min.
11. Place the slides in the machine or in the in situ slide adapter of a PCR machine, pipette 30 μL of hybridization solution per slide onto the metaphases and cover with plastic coverslips.
12. Run the following program o/n:

    78 °C for 10 min.

    50 °C for 1 min.

    45 °C for 90 s.

    40 °C for 2 min.

    38 °C for 5 min.

    37 °C for 16 h.
13. Proceed to Subheading 3.9.

3.6 Wax Embedding of Oat Roots for DNA and RNA FISH

1. Germinate oat seeds on filter paper and allow the roots to grow to a length of 2–3 cm.
2. Fix 5 mm root tips at room temperature for 4 h to o/n in 4% formaldehyde in PBS in 15 mL tubes. At the start apply a gentle vacuum three times for 5 min. This allows tissue to rise to the surface and then sink when the vacuum is turned off (*see* **Note 16**).
3. Wash root tips in PBS for 5 min to remove fixative.
4. Dehydrate root tips in an ethanol series (30%, 50%, 70%) for 30 min each. For thicker samples you may need to increase the time.
5. Place root tips in tissue or biopsy cassettes that are submersed in 70% ethanol to avoid drying of the material and quickly transfer the cassettes to a tissue processor for wax embedding.
6. Use the following settings with pressure and vacuum cycles for each step:

    70% ethanol for 1 h at 35 °C.

    80% ethanol for 1.5 h at 35 °C.

    90% ethanol for 2 h at 35 °C.

    Absolute ethanol for 1 h at 35 °C.

    Absolute ethanol for 1.5 h at 35 °C.

    Absolute ethanol for 2 h at 35 °C.

    Xylene for 30 min at 35 °C.

    Xylene for 1 h at 35 °C.

    Xylene for 1.5 h at 35 °C.

    Paramat wax for 1 h at 60 °C.

Paramat wax for 1 h at 60 °C.

Paramat wax for 2 h at 60 °C.

7. When the program has finished place the cassettes in the hot wax chamber of a paraffin wax embedding station. Place three parallel root tips in a metal mold on a thin layer of wax, cover with wax, add a mesh support and let the wax set (*see* **Note 17**).
8. Trim the block face to a trapezoid shape and cut 8–12 μm thin longitudinal sections on a rotary microtome.
9. Cover a 20 mm × 20 mm in the middle of a poly-L-lysine coated slide with deionized water, place the sections on the water and leave to dry on a slide warmer (*see* **Note 18**).
10. For DNA FISH proceed to Subheading 3.7. For RNA FISH proceed to Subheading 3.8.

3.7 Pretreatment and Hybridization of Root Sections for DNA FISH

1. Dewax sections in histolene twice for 30 min and rehydrate in an ethanol series (30 %, 50 %, 70 %, absolute ethanol), 90 s each step starting with absolute ethanol, before a 5 min wash in PBS.
2. Cover the sections with 100–200 μL of the enzyme mix and add a plastic coverslip. Incubate in a humid chamber for 1 h in an incubator at 37 °C. Wash 5 min in PBS.
3. Incubate the slides in a Coplin jar in 0.1 mg/mL RNase A in 2× SSC for 1 h at 37 °C in a water bath. Wash 10 min in PBS.
4. Dehydrate slides in an ethanol series, the reverse of above and leave to dry (*see* **Note 14**).
5. Prepare hybridization solution, 40 μL per slide, for a 20 mm × 20 mm area:

   20 μL formamide, deionized.

   8 μL 50 % dextran sulfate.

   4 μL 20× SSC.

   0.5 μL 10 % SDS.

   1 μL salmon sperm DNA.

   5 μL probe (200 ng of the nick translation Subheading 3.1 or deionized water as a no-probe control (*see* **Note 15**)).

   1.5 μL H_2O.
6. Denature the hybridization solution for 5 min in a water bath at 95 °C, cool on ice for 5 min.
7. Add water to the hybridization machine and place the slides in the machine or in the in situ slide adapter of a PCR machine. Evenly distribute the solution over the sections and add a plastic coverslip. Hybridize o/n using the following conditions:

   75 °C for 8 min.

   50 °C for 1 min.

45 °C for 90 s.

40 °C for 2 min.

38 °C for 5 min.

37 °C for 16 h.

8. Proceed to Subheading 3.9.

3.8 Pretreatment and Hybridization of Root Sections for RNA FISH

1. Dewax sections in histolene twice for 30 min and rehydrate in an ethanol series (30%, 50%, 70%, absolute ethanol), 90 s each step starting with absolute ethanol, before a 5 min wash in PBS (*see* **Note 19**).
2. Cover sections with 100 μL of the enzyme mix, cover with a plastic coverslip and incubate in a humid chamber for 1 h at 37 °C in an incubator. Wash 5 min in PBS.
3. Treat sections in a Coplin jar with 10 μg/mL proteinase K in proteinase K buffer for 30 min at 28 °C in a water bath. Prewarm buffer and container and add proteinase K just before use from a 10 mg/mL stock in deionized water (*see* **Note 20**).
4. Inactivate proteinase K for 5 min by incubation of slides in a solution of 2 mg/mL solution glycine in PBS at RT followed by a 5 min wash in PBS.
5. Postfix for 10 min in 4% formaldehyde in PBS, pH 7.4 in the fume hood and wash 2× 5 min in PBS.
6. Dehydrate in an ethanol series (the reverse of above) and leave to dry (*see* **Note 14**).
7. Prepare hybridization solution (40 μL/slide, for a 20 mm×20 mm area):

   4 μL formamide, deionized.

   1 μL tRNA.

   1–2 μL probe (ca. 200 ng of the in vitro transcription Subheading 3.2 or deionized water as a no-probe control (*see* **Note 15**)).

   *x*μL H_2O to a final volume of 8 μL.

   32 μL hybridization buffer (*see* **Note 21**).
8. Denature the hybridization solution at 80 °C for 2 min, cool on ice for 5 min.
9. Evenly distribute the solution over the sections and add a plastic coverslip, hybridize at 50 °C o/n in a humid chamber. Higher temperatures increase the stringency of the hybridization (*see* **Note 22**).
10. Proceed to Subheading 3.10.

3.9 Post-hybridization Washes for DNA FISH

1. Heat shaking water bath in the fume hood to 42 °C.
2. Wash slides in a Coplin jar for 3 min in 2× SSC in the shaking water bath at 42 °C to remove the coverslips. They will float and can easily be taken out.
3. Wash the slides 2× 5 min in 20 % formamide in 0.1× SSC in the shaking water bath at 42 °C.
4. Wash the slides 2× 5 min in 2× SSC in the water bath at 42 °C.
5. Wash the slides 2× 5 min in 2× SSC at RT on the shaker.
6. Wash the slides 2× 5 min in 4× SSC, 0.2 % (v/v) Tween 20 at RT.
7. Proceed to Subheading 3.11.

3.10 Post-hybridization Washes for RNA FISH

1. Heat shaking water bath in the fume hood to 50 °C.
2. Wash slides in a Coplin jar for 3 min in 2× SSC at 50 °C in the shaking water bath to wash off the coverslips. They will float and can easily be taken out.
3. Wash the slides for 15 min in 2× SSC/50 % formamide at 50 °C in the shaking water bath (*see* **Note 23**).
4. Wash the slides for 15 min in 1× SSC/50 % formamide at 50 °C in the shaking water bath.
5. Wash the slides for 5 min in 2× SSC at RT on a shaker.
6. Wash the slides 2× 5 min in 4× SSC, 0.2 % Tween 20 at RT.
7. Proceed to Subheading 3.11.

3.11 Immunodetection

1. Block the slides for 10 min in blocking solution, 100 μL/slide under plastic coverslips in a humid chamber.
2. Remove coverslips and tap solution off on paper.
3. For biotin labeled probes proceed to **step 5** only if no other label is present. Incubate the slides in mouse monoclonal anti digoxin antibody 1:5000 and/or rabbit anti-dinitrophenyl antibody 1:1000 in blocking solution (100 μL/slide) under plastic coverslips for 1 h at 37 °C in a humid chamber protected from light (*see* **Note 24**).
4. Wash the slides 3× 5 min in 4× SSC, 0.2 % Tween 20 in jar wrapped in tin foil on the shaker.
5. Incubate the slides in Alexa Fluor 488 goat anti mouse antibody 1:300 and ExtrAvidin-Cy3 1:300 and anti rabbit Alexa Fluor 647 antibody 1:600 in 5 % BSA/4× SSC/0.2 % Tween 20, 100 μL/slide, under plastic coverslips, 1 h at 37 °C in humid chamber in the dark (*see* **Note 25**).
6. Wash the slides 3× 5 min in 4× SSC/0.2 %Tween 20 in a jar wrapped in tin foil on the shaker at RT.

7. Counterstain in 100 μL 1 μg/mL DAPI with plastic coverslips in the dark for 3–4 min (*see* **Note 26**).
8. Wash briefly in 4× SSC/0.2 % Tween 20, tap excess solution off.
9. Mount slides in 15 μL of Vectashield or Prolong Diamond. Pipette the mounting medium onto the tissue/metaphases. Hold a no. 1.5 coverslip (22×22 mm) with forceps on one edge, lower the opposite edge onto the slide and then slowly lower the forceps until the coverslip sits on the sample. Mop up excess with filter paper and seal with nail polish (*see* **Note 27**).

4 Notes

1. You will need a good quality molecular grade DNase I, preferably from Roche, as cheap alternatives may contain traces of RNase.
2. Paraformaldehyde powder works equally well, but as it is a carcinogen, using prills, which are fine granules of paraformaldehyde, is a safer option avoiding the chemical being blown all over the fume hood.
3. Formamide is a skin irritant, may damage fertility or the unborn child and causes damage to organs through prolonged exposure! Use in the fume hood when dealing with more than μL quantities.
4. The concentrations of individual enzymes depend on the tissue. It is possible to start with 2 % cellulase "Onozuka" R10 (Yakult, Japan) on its own. In general probes will not penetrate any further than the first two cell layers even after weakening the cell walls.
5. Dextran sulfate is very viscous. When preparing the hybridization solution heat up an aliquot of the dextran sulfate to 65 °C and pipette with a cutoff pipette tip.
6. The stock solution can be defrosted a number of times. The working solution of 1 μg/mL DAPI can be kept in the fridge for several weeks. Protect from light.
7. Use gloves whilst preparing the plastic coverslips and sterilize the blades of your scissors with 70 % ethanol. These coverslips reduce evaporation and can easily be removed without damaging the sample.
8. Nick translation works for extracted genomic DNA, plasmids and PCR products. It is possible to directly incorporate fluorescently labeled nucleotides as an alternative to digoxigenin-, DNP- or biotin-labeled nucleotides that are then detected via immunolabeling. The former is recommended if there are persistent background problems. On the other hand immunolabeling

gives you valuable signal amplification. Reaction times can vary considerably. Keep the reaction on ice while running the gel and resume if necessary.

9. A negative control (usually the sense transcript) is necessary to check the specificity of probe binding. The T3 promoter added to the 5′ end of the forward primer can be used to synthesize the sense transcript in combination with the T3 polymerase. However, as many genes are controlled by noncoding RNAs it is possible that you see specific signal with the sense probe. An alternative control would be to use the transcript of a different gene with known RNA distribution.
10. Use sterile microfuge tubes and pipette tips.
11. Do NOT use hydrochloric acid to adjust the pH as formaldehyde reacts with HCl to produce the carcinogen Bis(chloromethyl) ether. Use pH strips to determine the pH rather than a pH electrode as fixatives can degrade pH electrodes. Formaldehyde solution has to be prepared on the day it is used.
12. Enzyme concentrations per manufacturer batch can vary. In 45–90 min the roots should be partially digested so that the tips dangle down when you tap the tip of the microfuge tube gently. If this is not the case, you will need to adjust the enzyme concentrations.
13. Slides with metaphase spreads can be stored for up to 2 weeks at 4 °C in the presence of a desiccant.
14. The slides need to be completely dry so that the hybridization solution is not diluted.
15. A no-probe control shows up unspecific binding of the primary antibody. The latter is usually due to insufficient washing after the antibody incubation or too high antibody concentrations.
16. Harvest your material speedily to avoid any tissue degradation before fixation. Vacuum infiltration allows the fixative to penetrate into air pockets, which is important for uniform tissue preservation. Only those samples, which have sunk to the bottom after the final vacuum application, will fix well. However, feathery tissue may never sink and still fix. If necessary, apply a further round of vacuum or cut smaller pieces. Some material such as leaf needs the addition of detergents to the fixative. A combination of Tween 20 and Triton X-100 both at 0.1 % works well.
17. Embedded material can be stored at 4 °C for several weeks.
18. Dry sections can be stored for several days at 4 °C.
19. You need a series of glass Coplin jars to pass your slides through. The jars should be dedicated to particular steps. At the very least always use the same jar for the proteinase treatment. From this stage onwards the RNA in your sample is vulnerable to attack by ubiquitous RNases. After the hybrid-

ization the RNA will be double-stranded and protected. We do not use DEPC-water but autoclave deionized water and buffers. Make sure that you wear gloves and that pipette tips etc. are sterile.

20. Mix in proteinase K with a sterile glass pipette to produce a homogeneous solution.
21. Warm up the hybridization buffer before use and pipette with a cutoff pipette tip.
22. 50 °C is the standard high stringency temperature used for hybridization. If you are not getting any labeling decrease the temperature to a minimum of 37 °C. If you suspect nonspecific binding increase the temperature.
23. As an alternative to **steps 3** and **4** you can wash in 0.2× SSC at 55 °C for 35 min followed by two washes in 0.2× SSC at 55 °C for 25 min each.
24. The protocol gives three different labels, which can be combined. For single color experiments use digoxigenin or DNP label for detection in tissue and any of the three for metaphase spreads. The detection of biotin label can lead to background caused by endogenous biotin in the tissue. Omitting the primary antibody will show if the secondary antibody binds nonspecifically.
25. The protocol gives a suitable fluorophore combination for triple labeling. For two-color labeling Alexa Fluor 488 and Alexa Fluor 568 are the preferred dyes conjugated to anti-mouse and anti-rabbit antibodies respectively. Extravidin-Cy3 is an alternative to Alexa Fluor 568. Alexa Fluor 647 is only used in triple labeling experiments as it is not visible to the naked eye. Once the fluorophores have been applied to the slides the latter need to be kept out of the light as far as possible to prevent bleaching.
26. A separate DAPI staining step is preferable to using a mounting medium that contains DAPI as you can ensure that your DAPI is fresh and that you have no unbound DAPI fluorescing in the background.
27. Do not use Vectashield with Alexa Fluor 647 because it reduces fluorescence of far red dyes.

Acknowledgments

I thank Ali Pendle for critical reading of the manuscript and the group of Enrico Coen for updating the RNA in situ methods. The work was supported by grants from BBSRC (BB/C504435/1, BB/J004588/1) and the John Innes Foundation.

References

1. Wegel E, Vallejos RH, Christou P et al (2005) Large-scale chromatin decondensation induced in a developmentally activated transgene locus. J Cell Sci 118(5):1021–1031

2. Wegel E, Koumproglou R, Shaw P et al (2009) Cell type-specific chromatin decondensation of a metabolic gene cluster in oats. Plant Cell 21(12):3926–3936

Chapter 2

Oat Doubled Haploids Following Maize Pollination

Philip A. Davies and Parminder K. Sidhu

Abstract

Doubled haploids (DHs) are an important tool for the accelerated production of new crop varieties. In oat, DHs were first produced by pollinating oat florets with maize pollen. The resultant embryos spontaneously eliminate the maize chromosomes leaving a haploid complement of oat chromosomes. These embryos can be cultured in vitro using the "embryo rescue" technique to produce haploid plants whose chromosome number can be doubled with colchicine to produce homozygous DH oat plants.

Key words Oat, Maize, Haploid, Doubled haploid, Breeding, *Avena sativa*, *Zea mays*

1 Introduction

Haploid plants can be produced from either male or female gametophytes which have a single copy of each chromosome pair. When the chromosome number is doubled either spontaneously or by treating plants with antimitotic agents like colchicine, the resultant doubled haploid (DH) plants have two identical sets of chromosomes, are completely homozygous and consequently true breeding. Achieving homozygosity in a single generation has the potential to reduce the time to breed a new variety by several generations in many self-pollinating cereal species [1]. Furthermore, the proportion of plants homozygous for genes of interest to breeders is much higher in a DH population compared to the F_2 and subsequent generations.

Oat DH plants can be produced from the male gametophyte using anther culture [2] or isolated microspore culture [3] DH oats can also be produced from the female gametophyte using embryo rescue following pollination with maize [4] as first developed for wheat DHs [5]. This chapter describes a method for generating oat DHs using oat × maize pollination, followed by dicamba spray, embryo rescue, and colchicine treatment.

Sebastian Gasparis (ed.), *Oat: Methods and Protocols*, Methods in Molecular Biology, vol. 1536,
DOI 10.1007/978-1-4939-6682-0_2, © Springer Science+Business Media LLC 2017

2 Materials

2.1 Donor Plant Growth Requirements

1. To achieve consistent results, it is best to grow both oat and maize donor plants in controlled environment rooms. Either sodium vapor or metal halide lamps at 400 $\mu E/m^2/s$ can be used, but we prefer sodium vapor lamps for growth of oat plants.
2. Plants are sown in 20 cm diameter pots with University of California potting mix containing slow release fertilizer such as Osmocote® (Scotts Australia, New South Wales, Australia) at 4 kg per 1000 kg soil and fertilized fortnightly with balanced liquid fertilizer such as Phostrogen (Debco, Melbourne, Australia) at the recommended rate (0.5 g Phostrogen per liter water; NPK ratio 14:4.4:22.5 with trace elements).
3. The University of California mix is prepared in bulk by mixing 1200 L of sterilized sand with 750 L of peatmoss with the addition of calcium hydroxide (hydrated lime, 1 kg) calcium carbonate (agricultural lime, 1.8 kg) and NPK fertilizer (Nitrophoska 12-5-14, 2 kg). The pH of this soil mix is 6.8.

2.2 Plant Germplasm

1. Any oat genotypes can be used but some genotypes are more amenable than others.
2. Most varieties of sweet corn can be used as pollinators but particularly good results can be obtained using the Mr Fothergills varieties Early Extra Sweet F1 or Kelvedon Glory F1.

2.3 Crossing Equipment and Post-crossing Treatment

1. Fine forceps and scissors for emasculation and clear glassine bags (approximately 10 cm × 25 cm) to cover the emasculated panicles.
2. A piece of paper and a small sable-hair paintbrush to collect and apply the maize pollen to emasculated florets.
3. A small water spray bottle for post-pollination growth regulator, dicamba (3,6-dichloro-*o*-anisic acid) application.
4. A dicamba solution (100 mg/L), prepared by weighing 100 mg dicamba and adding it to a 1 L bottle rinsed with ethanol. Dissolve the dicamba in 10 mL of ethanol. Make to a volume of 1 L with RO water.

2.4 Culture Medium Preparation

1. Gamborg's B5 basal media [6] with minimal organics (*see* **Note 1**).
   (a) *Solution A*: measure 500 mL ultrafiltered water into a beaker, add 20 g sucrose and stir until dissolved. Add 2.4 g of Gamborg's B5 basal media with minimal organics (Sigma) and stir until dissolved. Adjust pH to 6.0 with 1 M KOH.
   (b) *Solution B*: add 10 g agar (*see* **Note 1**) to a 1 L beaker containing 500 mL of distilled water and microwave for 5 min

on high setting. Remove from microwave and place on a hotplate with heat and stir with magnetic stirrer until agar is dissolved.

2. When solution B becomes clear, add solution A to it and mix thoroughly.
3. Pour approximately 10 mL of the mixed solution into each polycarbonate tube with flat base and autoclave the tubes for 15 min at 121 °C.
4. After autoclaving, slant the media in racks at a 45° angle to set as slopes at room temperature (*see* **Note 2**). Although fresh media gives best results it is possible to store media for up to 4 weeks at 4 °C.

2.5 Embryo Rescue and Culture

1. A laminar flow sterile work bench and a stereo dissecting microscope for aseptic dissection of embryos.
2. Ethanol (70%), bleach (1% active chlorine), sterile distilled water, and sterile McCartney jars are required for surface sterilization of caryopses.
3. Fine forceps, 90 mm sterile petri dishes are required for embryo rescue.
4. Sterilized tubes containing B5 culture medium are required to grow excised embryos.
5. Excised embryos require incubation in a controlled environment growth cabinet at 22–24 °C with a 16 h photoperiod and light intensity of 50 μE/m^2/s for 2–3 weeks.

2.6 Transplanting Haploids from Agar to Soil

1. Large tweezers to remove plants from agar (caution—avoid damage to the roots).
2. 75 mm pots packed into horticultural punnet trays (20 pots/tray) covered with plastic bags to maintain humidity.

2.7 Chromosome Doubling

1. A fume hood, glass beakers, an aeration pump and a colchicine solution with dimethyl sulfoxide (DMSO) are required for chromosome doubling.
2. 75 mm pots for repotting the colchicine treated plants.
3. Plastic boxes with drainage holes, half filled with soil for transferring growing DH plants.
4. Colchicine stock solution: prepare a 0.4% (w/v) solution by dissolving 2 g colchicine in 500 mL water in a fume hood and store at 4 °C (*see* **Note 2**).
5. To prepare the 0.2% colchicine solution for use, mix 40 mL colchicine stock with 40 mL water and add 1 mL DMSO.

3 Methods

3.1 Donor Plant Growth

1. Grow oat plants in controlled environment rooms at 17/14 °C day/night with a 14 h photoperiod at 400 μE/m^2/s in 20 cm diameter pots using University of California potting mix containing slow release fertilizer and fertilize fortnightly with balanced liquid fertilizer.
2. Grow maize plants at 22 °C with a 16 h photoperiod using the same pot size, potting soil, nutrition and light intensity as for oat plants.

3.2 Crossing

1. Emasculate oat by first removing secondary florets and then removing anthers from primary florets with fine forceps 1–2 days before anthesis. Care must be taken to minimize injury to the stigma. Cover the emasculated panicles with glassine bags. After 2–3 days, emasculated florets with feathery and fluffy stigmas are ready for pollination (*see* **Note 3**).
2. Discard old maize pollen by slightly shaking the tassels 30 min prior to pollen collection. Collect fresh pollen periodically at 30 min intervals by shaking pollen from tassels onto a sheet of paper (*see* **Note 4**).
3. Immediately prior to pollination, clip the oat glumes, lemma, and palea just above the stigma and apply fresh maize pollen to the stigma with a small sable-hair paintbrush. Then cover the pollinated panicle with a glassine bag to prevent unwanted cross-pollination.
4. The day after pollination, spray the panicle with an aqueous solution of dicamba (100 mg/L) (*see* **Note 2**). Repeat it the next day (*see* **Note 5**).
5. 13–15 days after pollination, remove the panicles from the plant by cutting the stem as low as possible and place the base of the panicle into a beaker of water (*see* **Note 6**).
6. It is preferable to excise and culture the embryos on the same day that panicles are harvested but panicles can be left for up to 7 days at 4 °C if necessary.

3.3 Embryo Rescue and Culture

1. Using fine forceps, remove the developing caryopses from florets into sterilized McCartney jars.
2. Surface-sterilize caryopses by placing them in 70% ethanol for 30 s followed by 10–15 mL bleach (1% active chlorine) for 5 min and then two rinses in sterile distilled water.
3. Using a dissecting microscope in a laminar flow hood, aseptically remove embryos from caryopses using fine forceps and place on B5 medium in tubes [6].

Fig. 1 Haploid oat plants germinated from embryos on B5 culture medium

4. Incubate the cultured embryos at 4 °C for 2 days followed by 2 days in the dark at room temperature. Then transfer to a growth cabinet (22–24 °C, 16 h photoperiod, 50 μE/m^2/s) for 2–3 weeks (Fig. 1).

3.4 Transplanting Haploids from Agar to Soil

1. When embryos have germinated and established roots, transplant seedlings to soil in 75 mm diameter pots and cover with clear plastic bags to maintain humidity (*see* **Note 7**). Grow haploid plants in a controlled environment room with 17/14 °C day/night temperatures and a 14 h photoperiod.
2. Remove plastic after 3–5 days and grow plants in controlled environment room till ready for colchicine treatment (usually 3 weeks after transplanting to soil).

3.5 Chromosome Doubling

1. When oat haploid plants are at the third tiller stage, take out plants from pots, wash roots to remove all soil, trim roots to 2 cm in length and trim one third from the top of the shoots.
2. Place trimmed plants in a 250 mL glass beaker containing the 0.2% colchicine solution plus DMSO and treat for 3 h in a fume hood with aeration of the solution provided by aeration pump at room temperature in the light (Fig. 2). Any light source will assist this process. Make sure roots are fully immersed in the colchicine solution.
3. Following colchicine treatment, wash plants in running water for 1 h and transplant to 75 mm pots for seed production.

3.6 Seed Production

1. Grow DH plants using 75 mm pots placed in plastic trays half filled with soil for good seed production. This provides surplus soil for root growth. The plastic trays must have drainage holes to allow excess water to drain (Fig. 3).
2. Maintain DH plants in a greenhouse at 21/16 °C day/night temperatures under full sunlight until harvested (Fig. 4).

Fig. 2 Trimmed oat haploid plants prepared to undergo colchicine treatment. Note aeration pump with tubing in the foreground

Fig. 3 Oat plants 1 week (*right*) and several weeks (*left*) after colchicine treatment. Note that the 75 mm pots are placed in orange trays which are half-filled with potting soil to allow additional root growth

Fig. 4 Oat doubled haploid plants at or near maturity

4 Notes

1. Gamborg's B5 basal salt mixture with minimal organics (catalog number G5893) and agar (catalog number A7049) can be purchased from Sigma-Aldrich.
2. Must wear personal protective equipment to cover mouth, eyes, and skin when handling colchicine, DMSO, and dicamba. Weigh colchicine and dicamba powders on a balance in a fume hood.
3. Pollinate when all the emasculated florets are fully receptive (usually 3–5 days after emasculation).
4. If pollen becomes sticky and dark yellow due to pollen death, discard it and collect fresh pollen.
5. Post pollination, two consecutive dicamba sprays at 24 and 48 h are essential for good caryopsis formation.
6. Restricting time of embryo excision to 13–15 days post pollination results in better germination frequency of embryos.
7. Culture media set at 45° angle is easy to scoop out of the tubes to take out the seedlings.

References

1. Kasha KJ, Maluszynski M (2003) Production of doubled haploids in crop plants. An introduction. In: Maluszynski M, Kasha KJ, Forster BP, Szarejko I (eds) Doubled haploid production in crop plants. A manual. Kluwer, Dordrecht, pp 1–4
2. Kiviharju E, Moisander S, Laurila J (2005) Improved green plant regeneration rates from oat anther culture and the agronomic performance of some DH lines. Plant Cell Tissue Organ Cult 81:1–9
3. Sidhu PK, Davies PA (2009) Regeneration of fertile green plants from oat isolated microspore culture. Plant Cell Rep 28:571–577
4. Sidhu PK, Howes NK, Aung T, Zwer PK, Davies PA (2006) Factors affecting oat haploid production following oat × maize hybridisation. Plant Breed 125:243–247
5. Laurie DA, O'Donoughue LS, Bennett MD (1990) Wheat × maize and other wide sexual hybrids: their potential for genetic manipulation and crop improvement. In: Gustafson JP (ed) Gene manipulation in plant improvement, vol 2. Plenum, New York, pp 95–126
6. Gamborg OL, Miller RA, Ojima K (1968) Nutrient requirements of suspension cultures of soybean root cells. Exp Cell Res 50:151–158

Chapter 3

Wide Hybridization Between Oat and Pearl Millet

Takayoshi Ishii

Abstract

Wide hybridization is a one of the important techniques in plant breeding. Oat (*Avena sativa* L.) and pearl millet (*Pennisetum glaucum* L.) belong to different subfamilies of Poaceae. In generally, such distant relative species show uniparental chromosome elimination after successful fertilization. However, all seven pearl millet chromosomes are retained beside the genome of oat during embryogenesis. Hybrid seedlings develop, but show necrosis after light irradiation. Here, a detailed protocol for wide hybridization between oat and pearl millet is described.

Key words Haploid, Uniparental chromosome elimination, Spike culture, Embryo culture

1 Introduction

Wide hybridization is a widely applied method in the plant breeding process. The hybrid plant is valuable to introduce new traits such as biotic resistance, abiotic stress resistance, and quality traits into cultivars [1, 2]. Wide hybridization is also used for the production of haploids to accelerate the plant breeding process. This application was pioneered by Kasha and Kao [3] who revealed chromosome elimination and haploid formation after pollination *Hordeum vulgare* with *H. bulbosum*. Meanwhile uniparental chromosome elimination has been reported in many species hybrids [4].

The successful production of haploid oat plants has been found in crosses between oat and maize (*Zea mays*), pearl millet (*Pennisetum glaucum*), and *Tripsacum dactyloides* [5–7]. Oat and maize or pearl millet belong to different subfamilies of the family Poaceae namely Pooideae and Panicoideae, respectively. Generally, complete chromosome elimination takes place in such distant hybrids. However, Riera-Lizarazu et al. [8] reported that a small number of the oat–maize hybrid plants maintained some chromosomes of maize in the back ground of oat. Later Ishii et al. [9] reported a full set of pearl millet chromosomes in oat–pearl millet hybrid embryos. Oat–pearl millet embryos grow into the small

Sebastian Gasparis (ed.), *Oat: Methods and Protocols*, Methods in Molecular Biology, vol. 1536,
DOI 10.1007/978-1-4939-6682-0_3,

true hybrid seedlings, but have a necrotic trait after light irradiation [5]. Despite co-expression of the centromere-specific histone H3 (CENH3) genes encoded by oat and pearl millet, only the oat-type CENH3 is incorporated into the centromeres of both species in hybrid embryos [10].

In future, oat–pearl millet hybrid embryos and plants could become a valuable material to understand the process of chromosome elimination better or to study the behavior of the C3 and C4 photosynthesis in hybrids. However, generation of hybrid embryos or plants requires several steps like, (1) growing healthy parent plants, (2) emasculation of oat plants and crossing with pearl millet, (3) embryo rescue tissue culture, and (4) characterization of hybrid plants.

2 Materials

2.1 Plant Material

1. Oat (*Avena sativa*; $2n=6x=42$) cultivar 'Best Enbaku' is used as a female parent and pearl millet (*Pennisetum glaucum*; $2n=2x=14$) cultivar 'Ugandi' as pollinator (*see* **Note 1**).

2.2 Spike Culture and Embryo Culture Medium

1. 1000× stock solution of 2,4-dichlorophenoxyacetic acid (2,4-D): 100 mg 2,4-D is dissolved with 1 mL (100 mg/mL) dimethyl sulfoxide (DMSO) at 50 °C. Stock solution can be stored at −20 °C (*see* **Note 2**). To prepare a 1× 2,4-D solution (100 mg/L) dissolve 1000× stock solution of 2,4-D with distilled water and store it at 4 °C in darkness (*see* **Note 3**).
2. Spike culture medium 1 (before crossing): dissolve 80 g sucrose with 2 L tap water and keep it at 4 °C.
3. Spike culture medium 2 (after crossing): dissolve 80 g sucrose and 2 mL 1000× stock 2,4-D solution with 2 L tap water and keep it at 4 °C in darkness.
4. 1 % sodium hypochlorite solution: dilute sodium hypochlorite solution with distilled water.
5. Embryo culture medium: To prepare 1 L medium dissolve 2.3 g Murashige and Skoog Plant Salt Mixture, 50 mg myo-inositol, 0.2 mg thiamine HCl, 50 mg L-alanine, 20 mg L-cysteine, 10 mg L-arginine, 10 mg L-leucine, 10 mg L-phenylalanine, 10 mg L-tyrosine, 30 g sucrose, and 8 g agar TC in distilled water and adjust a pH of 5.8 with KOH solution and autoclave. After autoclaving add 400 mg L-glutamine and use 20 mL of embryo culture medium per petri dish (Ø 9 cm). Prepare petri dishes under sterile conditions and after solidification of culture medium seal petri dishes with Parafilm. Store prepared petri dishes at 4 °C in darkness.

2.3 Cytological Analysis of Hybrids

1. Fixation solution for embryos and plants: mix ethanol and glacial acetic acid at a ratio of 3:1 (V/V) (*see* **Note 4**).
2. 90%, 70%, and 50% ethanol solutions: Mix ethanol and distilled water in proper concentration and store solutions at room temperature.
3. 45% acetic acid: Mix glacial acetic acid and distilled water and keep it at room temperature.
4. Enzyme solution: Dissolve 20 mg of pectolyase Y23 and 20 mg of cellulase "ONOZUKA" R-10 in 1 mL distilled water and keep it at −20 °C.
5. DNA extraction: ISOPLANT II Kit (NIPPON GENE) (*see* **Note 5**).
6. Klenow fragment.
7. 10 μM random primer (nonadeoxyribonucleotide mixture; pd (N)9).
8. 0.2 mM dNTP mix (minus dTTP).
9. 1 mM fluorescein-12dUTP.
10. TaKaRa *ExTaq* DNA polymerase (*see* **Note 6**).
11. 10 μM forward and reverse primers for pearl millet centromere specificrepeatsequences5′-CCGAAGCACGAGTTTTTCAT-3′ and 5′-TTGCACTATTTCGGGGTACC-3′.
12. 1 mM tetramethyl-rhodamine-5-dUTP.
13. DNA precipitation solution: 3 M sodium acetate pH 5.2 and 2-propanol.
14. Chromosomal DNA denaturation solution for GISH and FISH: 0.2 M NaOH in 70% ethanol.
15. 2× saline sodium citrate buffer (2× SSC): 300 mM NaCl, 30 mM sodium citrate, pH 7.0.
16. FISH and GISH hybridization solution for one slide (10 μl): 50% formamide, 10% dextran sulfate, 50–100 ng labeled probe DNA, and fragmented 100 ng wheat rDNA [11] (for blocking of ribosomal DNA sequence) in 2× SSC buffer.
17. Slide washing buffer: 0.1% Triton X-100 in 2× SSC buffer keep at room temperature.
18. DNA staining solution: Vectashield (Vector Laboratories) containing 1 ng/μL 4′,6-diamidino-2-phenylindole (DAPI) keep at 4 °C in dark.

2.4 PCR Analysis of Hybrids

1. DNA extraction: DNeasy Plant Mini Kit (*see* **Note 7**).
2. PCR: TaKaRa *ExTaq* DNA polymerase.
3. Primer sets: A total of 23 SSR markers (PSMP, 2006, 2008, 2043, 2056, 2059, 2078, 2084, 2090, 2227, 2231, 2233,

2237, 2246, 2248, 2251, 2255, 2263, 2266, 2267, 2270, 2271, 2273, and 2274) [12, 13] and five STS markers (PSM, 305, 345, 716, 737, and 870; Millet Gene) [14] are used to detect seven pearl millet linkage groups (Table 1).

3 Methods

3.1 Plant Growth Condition

1. Grow oat plants in Ø18 cm pots sequentially at 15 °C day/5 °C night 10 h day time for 4 weeks, 20 °C day/15 °C night 10 h light time for 4 weeks and 25 °C day/20 °C night 14 h light time until flowering. Pearl millet plants are grown in Ø27 cm pots at 25–30 °C day/20 °C night 16 h light time until flowering (*see* **Note 8**).

3.2 Crossing Between Oat and Pearl Millet

1. Cut 5 cm below the second node from the flag leaf if panicles have emerged few spikelets from the flag leaf of oat. Use 15–20 spikelets around the middle part of the panicle. Remove other spikelets with scissors.
2. Cut off the upper half of the spikelet with scissors and use 1st floret of spikelet for emasculation with tweezers (Fig. 1). Cover panicles with a glassine paper bag after emasculation (*see* **Note 9**).
3. Place up to 15 culms in a 500 mL plastic bottle filled with spike culture medium.
4. Cover the culms with a plastic bag to prevent drying out. Place bottle in a 4 °C water bath under continuous light (around 3000 Lux) and the water bath containing the bottle in an incubator adjusted at 23 °C.
5. Three days after emasculation, pollinate the florets with fresh pearl millet pollen and cover it with a plastic bag to prevent drying out (*see* **Note 10**). Keep the culms under the same condition as previous for 12 h.
6. Twelve hours after pollination drop onto each floret around 50 μL 1× 2,4-D solution using a syringe.
7. Cut the panicle 15 cm below the first node from the flag leaf and culture in a 500 mL plastic bottle containing spike culture medium 2. Keep the culms under the same condition as previous.
8. Renew every 3 days the spike culture medium 2 and wash the cutting edge of the culms under running tap water before placing it in fresh solution.

3.3 Embryo Culture

1. Sterilize 14 days after pollination immature seeds with 70% ethanol for 1 min and with 1% of sodium hypochlorite solution for 10 min.

Table 1
Primer sets to detect the seven different linkage groups of pearl millet

Marker	Linkage group	5′–3′ sequence of primers (foward_reverse)	References
PSMP2006	1	GACTTATAGTCACTGGGAAAGCTC_GCTTTAATAACTTTGTGCGTATT	[12]
PSMP2090	1	AGCAGCCCAGTAATACCTCAGCTC_AGCCCTAGCGCACAACACAAACTC	[12]
PSMP2246	1	CGGATGCTAAATTAACCGAAGC_CCAGCTTGCTTCTGTTGCGTTC	[13]
PSMP2273	1	AACCCCACCAGTAAGTTGTGCTGC_GATGACGACAAGACCTTCTCTCC	[13]
PSMP2059	2	GGGGAGATGAGAAAACACAATCAC_TCGAGAGAGGAACCTGATCCTAA	[12]
PSMP2231	2	TTGCCTGAAGACGTGCAATCGTCC_CTTAATGCGTCTAGAGAGTTAAGTTG	[13]
PSMP2237	2	TGGCCTTGGCCTTTCCACGCTT_CAATCAGTCCGTAGTCCACACCCCA	[13]
PSMP2255	2	CATCTAAACACAACCAATCTTGAAC_TGGCACTCTTAAATTGACGCAT	[13]
PSMP2056	3	ACCTGTAGCTTCAAAATTCAAAAA_AATTCAGTGTGATTTCGATGTTGC	[12]
PSMP2227	3	ACACCAAACACCAACCATAAAG_TCGTCAGCAATCACTAATGACC	[13]
PSMP2251	3	TCAAACATAGATATGCCGTGCCTCC_CAGCAAGTCGTGAGGTTCGGATA	[13]
PSMP2267	3	GGAAGGCGTAGGGATCAATCTCAC_ATCCACCCGACGAAGGAAACGA	[13]
PSM305	4	TTGTCAGATTAATGTAATTATGCTAGG_TTGCTATGTGATATATTGTATGTCTAGATG	[14]
PSM716	4	ACTTGCCGATCCAACTAACG_TTGAAAAACCTTCGATTCCG	[14]
PSMP2008	4	GATCATGTTGTCATGAATCACC_ACACTACACCTACATACGCTCC	[12]
PSMP2084	4	AATCTAGTGATCTAGTGTGCTTCC_GGTTAGTTTGTTTGAGGCAAATGC	[12]
PSM345	5	CTGGGGGAGAGAGAAGGG_AAAAGGCTGGGAGAGTAGGC	[14]
PSMP2078	5	CATGCCCATGACAGTATCTTAAT_ACTGTTCGGTTCCAAAATACTT	[12]

(continued)

Table 1
(continued)

Marker	Linkage group	5′–3′ sequence of primers (foward_reverse)	References
PSMP2233	5	TGTTTTCTCCTCTTAGGCTTCGTTC_ACCTTCTCCGCCACTAAACAACT	[13]
PSMP2274	5	CACCTAGACTCTACACAATGCAAC_AATATCAAGTGATCCACCTCCCAA	[13]
PSM737	6	ATGCTTTTCCCCCGCTATCA_TCCTAGCAAGCTCGCATAAGA	[14]
PSM870	6	TGGAACATCTGAAGTGCCTCA_GGAGCTAAGCATAGAAGCAGCA	[14]
PSMP2248	6	TCTGTTTGTTTGGGTCAGGTCCTTC_CGAATACGTATGGAGAACTGCGCATC	[13]
PSMP2270	6	AACCAGAGAAGTACATGGCCCG_CGACGAACAAATTAAGGCTCTC	[13]
PSMP2043	7	TCATATTCTCCTGTCTAAAACGT_CACAAATCGTACAAGTTCCACTC	[12]
PSMP2263	7	AAAGTGAATACGATACAGGAGCTGAG_CATTTCAGCCGTTAAGTGAGACAA	[13]
PSMP2266	7	CAAGGATGGCTGAAGGGCTATG_TTTCCAGCCCACACCAGTAATC	[13]
PSMP2271	7	CCTTATATTGGACCGACTGCTGAC_CTCCCCCATACACGAGCGAGAA	[13]

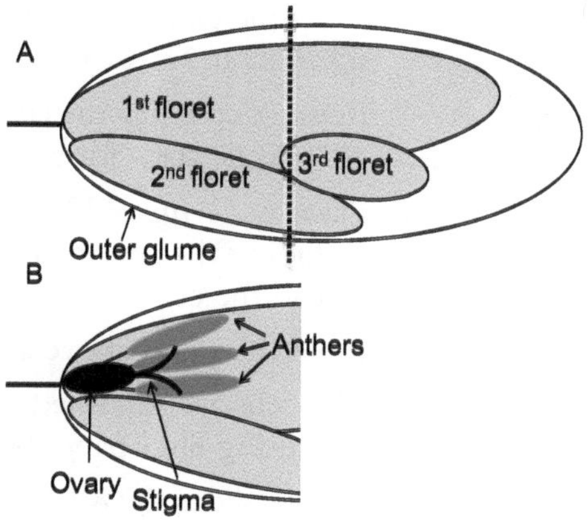

Fig. 1 Procedure of oat spikelet emasculation. (**a**) Each oat spikelet is composed with three florets. Cut spikelet with scissors on dash line position. (**b**) Remove 3rd floret with tweezers. Remove ovary and stigma with tweezers from 2nd floret. Remove three anthers from 1st floret with tweezers

2. Then rinse the immature seeds three times in distilled water for 1 min each time.
3. Excise embryos with tweezers under a stereo microscope, and place the scutellum side down on the embryo culture medium under sterile conditions.
4. Incubate petri dishes containing the embryos in the dark at 20 °C.
5. Once shoots develop, transfer the dishes to a growth chamber having 12 h light (2800 Lux)/12 h dark at a constant temperature of 20 °C (*see* **Note 11**).
6. 2–3 weeks later transfer the seedlings to a plant culture box filled with embryo culture medium and keep it under the same conditions for 2–3 weeks.
7. Transfer the plant to the soil and cover it with plastic bag for few days to prevent drying out.

3.4 Slide Preparation

1. Florets or plants are fixed with 10 mL of fixation solution at room temperature for 7 days (*see* **Note 12**).
2. To excise the embryos from florets or to excise plant meristematic parts, soak the florets and plants in 50% ethanol in a petri dish at room temperature for 1 h.
3. Dissect with tweezers under a stereo microscope.
4. Place embryos or plant meristematic parts onto a slide glass.
5. Wash the material with one drop of distilled water for 5 min at room temperature, then remove water carefully.

6. Treat the embryos or plant meristematic parts with an enzyme solution at 37 °C in humid condition.
7. Remove carefully the enzyme solution, and add a drop of 15 μL 45 % acetic acid (for enzyme treatment time *see* **Note 13**).
8. Place a 22 mm × 22 mm coverslip over the acetic acid drop including the specimen and gently tap with a woody needle to spread the cells.
9. Hold the slides above a spirit lamp flame for a few seconds.
10. The cells are squashed by carefully applying pressure with fingers using filter paper between coverslip and fingers.
11. Freeze slides at −80 °C overnight, then remove the coverslip with a razor blade.
12. Finally the slides are air dried for a few hours at room temperature (*see* **Note 14**).

3.5 DNA Labeling for GISH and FISH

1. Pearl millet total genomic DNA is extracted with an ISOPLANT II Kit following kit protocol.
2. Check the genomic DNA concentration with a NanoDrop and run a 1 % gel to check the quality of DNA (*see* **Note 15**).
3. Genomic DNA fragments with a size of around 300 bp should be prepared by sonication or autoclaving.
4. Label pearl millet genomic DNA (as GISH probe) with fluorescein-12dUTP by the Random Primer DNA Labeling method. Mix 30 μL of distilled water and 3 μL of pearl millet fragmented genomic DNA in a 1.5 mL tube.
5. Boil the closed tube for 5 min in water, and then chill the tube on ice for 5 min.
6. Add 5 μL of 10× Klenow fragment buffer, 5 μL of dNTP Mix (minus dTTP), 5 μL of random primers mix (10 μM), 1 μL of Klenow fragment (5 Unit) and 1 μL of fluorescein-12dUTP in the tube mix and then incubate at 37 °C overnight.
7. After incubation, add 5 μL of 3 M sodium acetate and 50 μL of 2-propanol to the tube and mix well, and then precipitate the labeled DNA at −20 °C for 1 h.
8. Centrifuge the sample at maximum speed at 4 °C for 30 min, remove the supernatant and wash the DNA pellet with 70 % ethanol 2 times.
9. Dry the pellet for few minutes at 37 °C, and dissolve it in 25 μL of distilled water. Store the probe at −20 °C.
10. Label pearl millet centromere-specific tandem repeats with tetramethyl-rhodamine-5-dUTP by using the PCR method. Therefore mix 36.8 μL distilled water, 5 μL 10× Taq buffer, 4 μL dNTP Mix (minus dTTP), 1 μL of each forward and

reverse primer (10 μM), 1 μL pearl millet genomic DNA (50 ng/μL), 1 μL tetramethyl-rhodamine-5-dUTP and 0.2 μL Takara *ExTaq* DNA polymerase (5 U/μL) in a PCR tube.

11. Reaction is run with an initial denaturation step of 5 min at 94 °C; 35 cycles of 94 °C for 30 s, 57 °C for 30 s, and 72 °C for 0 s; and a final extension for 7 min at 72 °C (*see* **Note 16**).
12. PCR product is precipitated by the same method as the random primed labeled DNA and dissolved in 25 μL of distilled water. Store the probe at −20 °C.

3.6 In Situ Hybridization

1. Denature chromosomal DNA with the chromosomal DNA denaturation solution for 5 min at room temperature.
2. Dehydrate sequentially in a series of 70%, 90% and 99.6% ethanol at room temperature.
3. Air-dry the slides at room temperature for a few minutes.
4. Denature hybridization solution in boiling water for 5 min, and then chill it on ice water for 5 min.
5. Apply 10 μL of hybridization solution per air-dried slide, and cover each slide with a 22 mm×22 mm coverslip.
6. After glue sealing incubate the slides at 37 °C overnight under humid condition.
7. Remove the coverslips, wash the slides with the slide washing buffer for 5 min and next in 2× SSC buffer for 5 min at room temperature.
8. Dehydrate the slide sequentially in 70%, 90% and 99.6% ethanol at room temperature, and then air dry at room temperature for a few minutes.
9. Counterstain the slides with 8 μL of nucleic staining solution and cover with 22 mm×22 mm coverslips.
10. Capture images with a cooled CCD camera using a fluorescence microscope (Fig. 2) (*see* **Note 17**).

3.7 PCR-Based Genotyping for Plants

1. Extract DNA of seedlings with the DNeasy Plant Mini Kit following the kit protocol.
2. Perform the PCR in a total volume of 50 μL containing 50 ng template DNA, 0.2 μL TaKaRa *ExTaq* DNA polymerase (5 U/μL), 5 μL 10× *ExTaq* buffer, 1 μL each of forward and reverse primer (10 μM), and 4 μL dNTP mixture (2.5 mM each).
3. Run reactions with an initial denaturation step of 5 min at 94 °C; 35 cycles of 94 °C for 30 s, 57 °C for 30 s, and 72 °C for 30 s; and a final extension for 7 min at 72 °C.
4. Gel separate the amplicons using a 3% agarose gel followed by staining with ethidium bromide to visualize the bands (Fig. 3) (*see* **Note 18**).

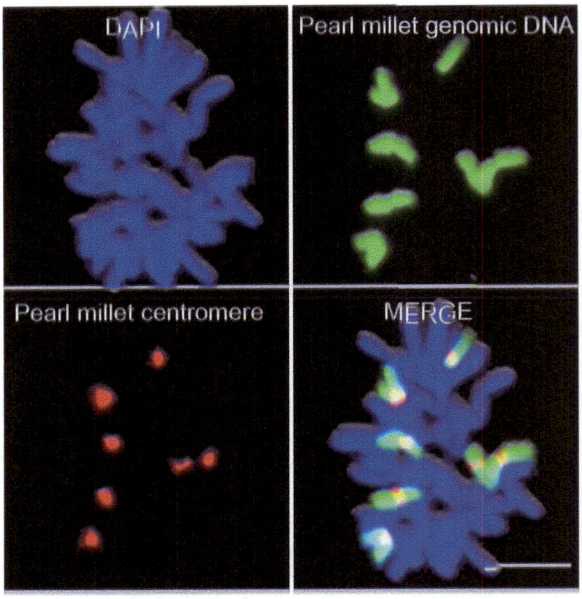

Fig. 2 FISH analysis of oat–pearl millet hybrids. Oat–pearl millet prometaphase cell contains all seven pearl millet chromosomes in the background of oat. Pearl millet-specific GISH signals in *green*, centromere-specific signals of pearl millet are *red*. Chromosomes counterstained with DAPI are *blue*. Scale bar: 10 μm

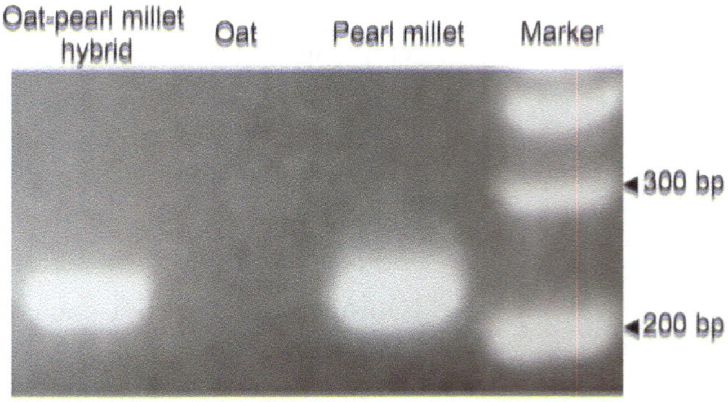

Fig. 3 PCR with genomic DNA of an oat–pearl millet hybrid, oat and pearl millet using pearl millet specific PCR primers. The oat–pearl millet hybrid sample shows the same PCR product as the pearl millet sample

4 Notes

1. Embryo formation rate differs between different crossing combinations of cultivars. The expected embryo formation rate is around 3 %.
2. The stock solution should not be stored longer than 1 year at −20 °C.

3. 50 mL of working solution is preferable. Do not store the 1× 2,4-D solution longer 1 month at 4 °C in dark.
4. Always prepare a fresh fixation solution.
5. Pearl millet contains a lot of polysaccharide, which very difficult get rid during DNA extraction procedure. DNA purity and concentration (1 μg/μL) are crucial point to generate probe.
6. Use proof reading DNA polymerase to incorporate a tetramethyl-rhodamine-5-dUTP to the PCR products.
7. Minimum DNA concentration is around 50–100 ng/μL.
8. Oat and pearl millet require a good light source to grow (more than 30,000 Lux). Oat needs around 5–6 months until flowering. Pearl millet needs around 4–5 months until flowering.
9. Timing of emasculation is different for different cultivars of oat.
10. Collect fresh pearl millet pollen (around 9–10 AM), and distribute pearl millet pollen with a small brush to the oat stigmas.
11. The combination oat×pearl millet results in a high frequency of true hybrids. True hybrids show necrosis after light irradiation. However, it is possible to keep hybrid plants under dark conditions for a few months without necrosis. In contrast, haploid oat plants are able to survive the light irradiation.
12. Fixation 7 and 14 days after pollination of florets is preferable. It is possible to keep fixed material in 70% ethanol at 4 °C for long storage.
13. The enzyme treatment time is crucial. Time of treatment depends on the sample size and age of the embryos. Treat for 5–10 min 7 day old embryos (300–500 cells). Treat for 45–60 min 14 day old embryos (more than 5000 cells). Treat plant meristematic parts for around 30 min.
14. Mark the position of the sample on the slide with a diamond pen. It is possible to store the slides in −80 °C until use.
15. High quality genomic DNA is crucial for efficient DNA labeling. Adjust final genomic DNA concentration of 1 μg/μL with distilled water.
16. PCR product size is depends on the primer sets.
17. In true hybrids up to 7 centromere signals are detectable after FISH.
18. In true hybrids up to 7 linkage groups are detectable after PCR.

Acknowledgement

I would like to thank Dr. Andreas Houben, IPK, Germany, for reading and providing valuable comments on this manuscript.

References

1. Dwivedi SL, Stalker HT, Blair MW, Bertoli DJ, Nielen S, Ortiz R (2008) Enhancing crop gene pools with beneficial traits using wild relatives. Plant Breed Rev 30:179–230
2. Kaneko Y, Bang SW (2014) Interspecific and intergeneric hybridization and chromosomal engineering of Brassicaceae crops. Breed Sci 64(1):14–22
3. Kasha KJ, Kao KN (1970) High frequency haploid production in barley (Hordeum vulgare L.). Nature 225(5235):874–876
4. Ishii T, Karimi-Ashtiyani R, Houben A (2016) Haploidization via chromosome elimination: means and mechanisms. Annu Rev Plant Biol 67:421–38
5. Rines HW, Dahleen LS (1990) Haploid oat plants produced by application of maize pollen to emasculated oat florets. Crop Sci 30(5):1073–1078
6. Matzk F (1996) Hybrids of crosses between oat and andropogoneae or paniceae species. Crop Sci 36(1):17–21
7. Ishii T, Tanaka H, Eltayeb AE, Tsujimoto H (2013) Wide hybridization between oat and pearl millet belonging to different subfamilies of Poaceae. Plant Reprod 26(1):25–32
8. Riera-Lizarazu O, Rines HW, Phillips RL (1996) Cytological and molecular characterization of oat x maize partial hybrids. Theor Appl Genet 93(1-2):123–135
9. Ishii T, Ueda T, Tanaka H, Tsujimoto H (2010) Chromosome elimination by wide hybridization between Triticeae or oat plant and pearl millet: pearl millet chromosome dynamics in hybrid embryo cells. Chromosome Res 18(7):821–831
10. Ishii T, Sunamura N, Matsumoto A, Eltayeb AE, Tsujimoto H (2015) Preferential recruitment of the maternal centromere-specific histone H3 (CENH3) in oat (Avena sativa L.) x pearl millet (Pennisetum glaucum L.) hybrid embryos. Chromosome Res 23(4):709–718
11. Gerlach WL, Bedbrook JR (1979) Cloning and characterization of ribosomal-RNA genes from wheat and barley. Nucleic Acids Res 7(7):1869–1885
12. Allouis S, Qi X, Lindup S, Gale MD, Devos KM (2001) Construction of a BAC library of pearl millet, Pennisetum glaucum. Theor Appl Genet 102(8):1200–1205
13. Qi X, Pittaway TS, Lindup S, Liu H, Waterman E, Padi FK et al (2004) An integrated genetic map and a new set of simple sequence repeat markers for pearl millet, Pennisetum glaucum. Theor Appl Genet 109(7):1485–1493
14. http://jicbio.bbsrc.ac.uk/cereals/millet.html

Chapter 4

Protocol for Producing Synthetic Polyploid Oats

Toshinobu Morikawa

Abstract

A protocol to produce synthetic oats by introducing alien genetic variations into cultivated oats is proposed based on suitable genetic relationships between *Avena* species. Furthermore, artificial hybridization procedures between different ploidy species are explained. Amphiploids can be produced by rescuing aborted embryos and treating the F_1 hybrids with colchicine to overcome sterility between interspecific plants. Furthermore, I present the cytological methods for observing somatic and meiotic chromosomes in the treated hybrids.

Key words *Avena sativa*, Polyploid, Amphiploid, Synthetic oats, Hybridization, Emasculation, Pollination, Embryo rescue, Chromosome doubling, Colchicine

1 Introduction

A synthetic allopolyploid is a plant that is grown from an experimental cross of two or more species or genera containing one or more copies of the parental genome. Synthetic hexaploid wheat has various improved traits, and its genes have been identified, mapped, and transferred to common wheat [1]. However, different from synthetic wheat, hybrid species and synthetic species of *Avena* are not suitable in the production of new cultivars. This agrees with the results of our investigations on a range of amphiploids synthesized in cereal crop species other than *Avena*.

However, synthetic polyploid oats contain suitable genetic material that can be utilized in studying genomic affinity among wild and cultivated species, translocation mechanisms among different genomes, transfer of disease resistance to new varieties, and polyploid evolution. Species hybrids and synthetic amphiploids of *Avena* have been utilized in transferring genetic variation from different ploidies of wild species into cultivated hexaploid oats [2]. Genomic differentiation among the *Avena* species has resulted in severely reduced chromosome homology, chromosome pairing, and gene exchange either during direct interploidy crossing or

Sebastian Gasparis (ed.), *Oat: Methods and Protocols*, Methods in Molecular Biology, vol. 1536,
DOI 10.1007/978-1-4939-6682-0_4, © Springer Science+Business Media LLC 2017

when combining with *A. sativa* and bridging types, such as synthetic hexaploids. Dilkova et al. [3] investigated C-banded karyotypes of somatic chromosomes and meiotic abnormalities in crown rust resistant lines derived from the synthetic hexaploid *A. abyssinica* × *A. strigosa* and selected the lines with normal performance for use as breeding stock. Ueno and Morikawa [4] reported the identification of C-genome and satellite (SAT) chromosome rearrangements in three synthetic hexaploid oats of *A. magna*, *A. murphyi*, and *A. barbata* × *A. strigosa* using genomic in situ hybridization. They performed fluorescence in situ hybridization of 18S-5.8S-26S rDNA and used this as a probe to detect the SAT chromosome carrying an intergenomic translocation segment. They also discussed the evolutionary occurrence of intergenomic translocations during formation of allopolyploids.

In this chapter, I describe a protocol to produce synthetic oats and propose procedures to introduce alien genes into cultivated oats based on suitable genetic relationships between *Avena* species. To overcome the sterility of the F_1 hybrids, amphiploids can be produced by rescuing aborted embryos and by treating them with colchicine.

2 Materials

2.1 Avena Gene Pools to Prepare a Synthetic Polyploid and Gene Transfer

Avena L. (Poaceae) comprises a polyploid series of diploid, tetraploid, and hexaploid species with a basic chromosome number of seven. Four genomes (A, B, C, and D) have been identified based on comparative karyotyping [5] and chromosome pairing between interspecific hybrids [6, 7]. These genomes evolved primarily through allopolyploidization of hybrids between wild diploid species. Common oat, *A. sativa* L., is a major cereal crop in traditional Old World agriculture and a close companion of wheat and barley. All *Avena* species can be crossed with each other within the primary gene pool of the same genomic species, but the hybrids tend to have chromosomal abnormalities and show sterility. Tissue culture techniques may be needed to rescue hybrid embryos from the secondary gene pool between different ploidy levels or genomes.

1. Diploid species: The evolution of diploid cultivated sand oats is similar to that of barley. Crops that vary in grain size and shape were divided into *A. strigosa*, *A. nuda*, *A. brevis*, and *A. hispanica*. They share the A-genome, which has the most variation among the four *Avena* genomes. The A-genome species include wild diploids, such as *A. prostrata*, *A. longiglumis*, *A. atlantica*, *A. canariensis*, and *A. damascena*. The C-genome diploid oats are all wild species, such as *A. clauda* and *A. eriantha*, which share the same genome and can interbreed. Nowadays, the B- and D-genome diploid species have not been observed.

2. Tetraploid species: Tetraploid oats consist of wild, weed, and cultivated species. A cultivated tetraploid (AABB), namely, *A. abyssinica* originated in Abyssinia, Ethiopia from the wild species *A. barbata* or from the weed species *A. vaviloviana*. The Abyssinian oat species is a very primitive cultivated form with limited genetic diversity to be incorporated into an elite oat variety. New tetraploid species, such as *A. magna*, *A. murthyi*, and *A. insulalis*, were discovered in a Mediterranean country. They include the A- or D-, and C-genomes and are becoming suitable candidates in the production of stable synthetic hexaploid oat with *A. strigosa*. The tetraploid wild species, *A. agadiriana* (AABB), was also newly found along the Mediterranean coast of Morocco.
3. Hexaploid species: Cultivated hexaploid oat involves *A. sativa* and *A. byzantina*. They include the A-, C-, and D-genomes. Common oat (*A. sativa*) is the main oat crop, which was domesticated in southwest Asia. According to a previous study, red oat (*A. byzantina*) is domesticated in northwest Africa [8]. In comparison with domestication of wheat and barley, the oat crop evolved from the hexaploid weed species *A. sterilis* or *A. fatua* through mimicry of vegetative plant tissue and seeds from wheat and barley cultivation. The hexaploid oat is interfertile.

2.2 Equipment for Producing Oat Hybrids

1. Small scissors.
2. Sharp-pointed forceps.
3. Glassine bags with natural starch (*see* **Note 1**).
4. Paper pins or clips.
5. Small tags.
6. 70% ethanol (*see* **Note 2**).

2.3 Gamborg Medium (B5) Without Plant Hormones to Rescue Embryos

Gamborg medium without plant hormones: 2500 mg/L KNO_3, 150 mg/L $CaCl_2{\cdot}2H_2O$, 250 mg/L $MgSO_4{\cdot}7H_2O$, 134 mg/L $(NH_4)_2SO_4$, 150 mg/L $NaH_2PO_4{\cdot}H_2O$, 0.75 mg/L KI, 3.0 mg/L H_3BO_3, 10 mg/L $MnSO_4{\cdot}H_2O$, 2.0 mg/L $ZnSO_4{\cdot}7H_2O$, 0.25 mg/L $Na_2MoO_4{\cdot}2H_2O$, 0.025 mg/L $CuSO_4{\cdot}5H_2O$, 0.025 mg/L $CoCl_2{\cdot}6H_2O$, 43 mg/L ferric-EDTA, 2% sucrose, 100 mg/L inositol, 1.0 mg/L nicotinic acid, 1.0 mg/L pyridoxine HCL, and 10 mg/L thiamine HCl, adjust pH 5.5 (*see* **Note 3**).

3 Methods

3.1 Artificial Hybridization Between Different Ploidy Species

Artificial hybridization between different polyploid species of oat is performed by crossing of selected parents. Specific methods are available for crossing different polyploid species, which differ according to floral morphology, genetic sexual barriers, and environmental factors. The female parental flower plays a central role in hybridization but the success of the crossing program depends on the health, readiness to pollinate, maturity, and other factors of the male parental flower.

3.1.1 Flower Health and Hybridization

Similar to other small grain crops, oat is also easy to be crossed; good results can be obtained by careful attention to details. It is best to avoid field-wide crossing due to the difficulty of maintaining good environmental conditions in the field.

1. Different ploidy levels of oat can be crossed in a greenhouse or growth chamber. Environmental conditions should be those that are optimum for plant growth and development.
2. Crossing in the growth-chamber can be performed during any season under optimum conditions, whereas most greenhouse crossings are made during the spring months.
3. In high latitudes, entire day even when the weather is clear, artificial lights, illuminated with metal halide light, provide a 13-h photoperiod (*see* **Note 4**).
4. Plants for greenhouse crossing can be grown in pots, which are often preferred because they can be easily moved when crossing.

3.1.2 Synchronization of Flowering

In artificial pollination, the breeder should be familiar with the species regarding its flowering habits, such as time from planting to flowering, duration of flowering, mechanisms of natural pollen dehiscence and fertilization, and the time of peak pollen production to take advantage of the window of opportunity of anthesis for the best artificial pollination outcomes [9].

1. Staggered planting is method in which the parent plants are planted at different times to ensure that the parents used in the crossing program will have flowers at the same time.
2. A long-day-photoperiod should be maintained in photoperiod-sensitive species to advance flowering as appropriate.
3. Other techniques that have been used in specific cases include delayed planting of early flowering species, manipulation of temperature and plant density, removing older flowers to induce a new flush of flowers, and pinching.

3.1.3 Emasculation

The female parent usually needs some special preparation. The flowers of the parent selected to be female are prepared for hybridization by removing the anthers, which is called emasculation. Pollen is often physically or manually transferred [10].

1. Usually ten primary and secondary floret tips near the top of the panicle are cut with small scissors to make it easier to remove three anthers in a floret with forceps when emasculating and pollinating.

2. Natural anthesis begins in the uppermost florets of a panicle and proceeds in the order of their emergence from the boot. Emasculation can be performed at any time of day except during natural anthesis.
3. A cross using *A. barbata* as a male parent is difficult due to the small size of the anthers and the occurrence of anthesis at midnight.

3.1.4 Pollination

1. Ideal conditions for pollination are clear days with moderate to low temperature. Most breeders allow 1–3 days between emasculation and pollination.
2. Natural oat anthesis in the field usually occurs in the afternoon, whereas greenhouse pollination is usually performed during late morning or early afternoon (*see* **Note 5**). A mature anther held with forceps is gently tapped against the inside wall of the floret for pollination where the lemma and palea are located (*see* **Note 6**).
3. After all florets have been pollinated, the glassine bag is replaced to prevent outcrossing (*see* **Note 7**). Watering the female plants carefully after pollination will lead to good results.

3.1.5 Embryo Rescue

A hybrid embryo that forms after fertilization of the crosses between different ploidy levels but fails to develop any further can be rescued. It is not always necessary to rescue embryos of hybrids between diploid and tetraploid species.

1. Breeders can intervene in the developmental process by dissecting the flower to remove the immature embryo. The embryo is then nurtured into a full plant using tissue culture technology.
2. The aseptic fertilized ovary is excised after removing the pericarp within 2 weeks of fertilization to avoid abortion. Normal embryogenesis ends at seed maturation. Prior to differentiation, the developing embryo is heterotrophic and dependent on the endosperm for nutrients.
3. Similar to all tissue culture work, embryo rescue is conducted aseptically and the embryos are cultured on B5 medium (pH 5.5) with 2 % sucrose and 0.2 % gellan gum (or 5 % agar) without auxin and cytokine (*see* **Note 8**).

3.1.6 Doubling Chromosomes of the Hybrids

Colchicine is applied to the meristematic region of the plant by wetting with an aqueous solution, spraying an emulsion, or rubbing on a lanolin paste [11].

1. Three-leaf stage oat hybrid plantlets are treated with 0.1 % colchicine aqueous solution, including 1 % dimethyl sulfoxide and a few drops of Tween 20 by submerging the plantlet's crown in a 10 cm long glass-tube for 24 h to double the number of chromosomes (*see* **Note 9**).

2. After washing in deionized water, the treated plantlets are transplanted to pots filled with a mixture of vermiculite and field soil.
3. The first generation plants (C_0) treated with colchicine are grown in a greenhouse and are self-pollinated to obtain the next generation seeds (C_1).
4. The polyploid sectors can be identified after inducing polyploidy in the diploids based on altered leaf morphology, fertility, larger cell size, and more chloroplasts in guard cells.
5. Often the polyploidy sector will develop tiller after treating the oat seedlings. The number of chromosomes must be determined to verify doubling. Adjacent diploid tillers or branches may need to be removed to reduce competition for the often slower developing polyploid tissues.

3.1.7 Counting Somatic Chromosomes in Oat Plants as per Darlington and La Cour's Method [12] with Some Modifications

The chromosome number of synthetic oats can be counted easily in the somatic complement. Roots (1–2 cm long) with their tips excised from seedlings germinated in Petri dishes usually give better results than those collected from potted plants.

1. Pretreatment: The root tips are pretreated with ice-cold water (0 °C) for 24 h or with 0.002 M 8-hydroxyquinoline for 2–3 h at 18 °C.
2. Fixation: Prior fixation in fresh Farmer's solution (ethyl alcohol–acetic acid = 3:1) for 12–24 h reduces staining of the cytoplasm.
3. Storage: If the material is used for immediate staining, it can be stored in 70% alcohol at 0–4 °C, but fixation and staining will suffer after storage for ≥3 months.
4. Staining: The pretreated and fixed roots are hydrolyzed in 1 N HCl acid for 12 min at 60 °C. After thorough rinsing in tap water, the hydrolyzed roots are placed in Feulgen's reagent (*see* **Note 10**) for a few minutes and then they are ready for squashes in 45% acetic acid or 2% aceto-carmine (*see* **Note 11**). The material can also be immersed in a small glass bottle containing aceto-carmine for 5–10 min.
5. Spreading: The cells are macerated by gently heating the bottle over an open flame. Afterwards, the cells are placed on a slide and covered using a coverslip. The prepared specimens are observed for the presence of prophase or metaphase chromosomes. The degree of spreading can be changed by applying pressure on the coverslip. Gentle finger-tip pressure on the coverslip under filter paper may assist in this process.

3.1.8 Aceto-carmine Squash Methods for Observing Meiotic Chromosomes in Oat Plants Following Rajhathy and Thomas [5] with Some Modifications

1. In oats, the developmental stage of inflorescence is most important for best sampling results. Although the best meiotic stage can vary depending on the growing conditions and genotype, the best sampling stage for meiosis is from the emergence of the flag leaf (2–3 cm).
2. The immature panicle is fixed in Carnoy's solution (6:3:1; alcohol–chloroform–acetic acid) for a few days (*see* **Note 12**) and stored in 70 % alcohol at 4 °C.
3. It is more difficult and time-consuming to locate anthers in the appropriate meiotic stage in the panicle than in a spike. In a spike, there is a sequence of stages occurring along the rachis of the spike, but the size of the spikelet in a panicle is the best guide to locate anthers.
4. Meiotic chromosomes in oat at metaphase tend to be sticky. Therefore, it is more difficult to obtain good chromosome spreads in oat than in other cereal species. The chromosome pairings observed at diakinesis are much better than those at metaphase particularly when multiple pairings are present. Stickiness can be reduced by storing longer in 70 % alcohol and by using alcoholic hydrochloric acid-carmine (*see* **Note 13**) [13] stain instead of aceto-carmine.

3.2 Examples of Synthetic Polyploid Oats

3.2.1 Synthetic Hexaploids

Synthetic 6× amphiploids have been produced by treating triploid F_1 hybrids between diploid and tetraploid species with colchicine. Due to meiotic irregularities and poor fertility, amphiploids are of limited value in the production of new species for cultivation. Good combinations of diploid and tetraploid amphiploid parents show stable chromosome pairings and seed fertility. The synthetic hexaploids (AsAsAACC), *A. strigosa* × *A. murphyi* and *A. strigosa* × *A. magna* suffice for the next generation [4].

3.2.2 Synthetic Octoploids

Synthetic octoploids combine the diploid genome with the sativa genome. Two combinations have been established, namely, *A. strigosa* × *A. sativa* (AsAsAACCDD) with resistance to crown rust [14] and *A. prostrata* × *A. sativa* (ApApAACCDD) with resistance to mildew [15]. The embryos in these cases needed to be rescued to obtain the hybrids. The level of fertility and stability attainted by the octoploid is insufficient to develop a cultivated octoploid species. However, synthetic octoploids have been used to bridge the barriers of infertility during the transfer of genetic variation to the cultivated forms. The specific synthetic octoploid, *A. sativa* cv. Pendek × *A. longiglumis* CW57 (A_lA_lAACCDD) has been established by homologous chromosome pairing [16].

3.2.3 Synthetic Decaploids

Pentaploid F_1 hybrids between natural tetraploid and hexaploid species are treated with colchicine to produce synthetic decaploids. To transfer the gene responsible for mildew resistance from the wild oat *A. barbata* into the cultivated oat, the synthetic decaploids (AsAsAABBCCDD) have been produced from the hybrid *A. barbata* × *A. sativa* [17]. As in the octoploid amphiploids, the A-genome is tetrasomic, but the low multivalent frequency shows that the A-genome pairs differently [2].

3.3 Future Prospects of Synthetic Polyploid Oats in Breeding

Synthetic polyploid oats are not suitable for the production of new cultivars. However, several attempts have produced synthetic species that can be used to improve oat characteristics.

In the past three decades, three new wild tetraploid species involving hexaploid oat progenitors were discovered in North Africa, Southern Spain, and Sicily, including *A. magna*, *A. murthyi*, and *A. insulalis* with the A- or D- and C-genomes. Ladizinsky [18] reported domestication by hybridizing the wild tetraploid oats *A. magna* and *A. muphyi*. He reported a new species, namely, *A. insulalis* from Sicily, which may be the tetraploid progenitor of hexaploid oats.

Ladizinsky [19] produced synthetic hexaploid oat by chromosome doubling of a sterile triploid hybrid between cultivated *A. strigosa* ($2n = 14$) cv. Saia and a domesticated form of *A. magna* ($2n = 28$). The synthetic hexaploid was intermediate between its parents in panicle shape and lemma color, similar to the tetraploid parent in spikelet structure and similar to the diploid parent having a single, albeit partially shriveled seed per spikelet and low protein content.

Rines et al. [20] adopted two methods to transfer the gene responsible for crown rust resistance from *A. strigosa* CI6954SP and *A. muphyi* P12 to a hexaploid oat cultivar. The first described method was a direct cross of the diploid to the hexaploid facilitated by rescuing embryos to produce a synthetic octoploid that allowed rapid recovery of fertility and the agronomic phenotype after subsequent backcrossing. The second method was crossing the diploid *A. strigosa* with the tetraploid *A. murphyi* followed by colchicine treatment to produce a synthetic hexaploid, which allowed transfer of two resistant genes to cultivated oat without rescuing the embryo.

Morikawa [21] reported transferring the dominant gene conferring mildew resistance from wild oat, *A. prostrata*, into cultivated oat, *A. sativa* cv. Sun II. The *prostrata* substitution line Av2557 was crossed with the synthetic octoploid, *A. sativa* cv. Pendek × *A. longiglumis* Cw57, and the hybrid was backcrossed to "Sun II." The Cw57 genotype of *A. longiglumis* increases homologous chromosome pairing and can be successfully used to induce disrupt control of chromosome pairing in *A. sativa*.

Use of a chromosome substitution line or addition line as a source of genotypes in a crossing program reduces the amount of screening required compared with direct crossing to the synthetic polyploids. In this way, much greater tolerance of *Avena* for genomic doubling will allow manipulations of resistance genes.

4 Notes

1. Transparent glassine bags are placed over the emasculated florets to protect them from out-crossing and drying out. Although various sizes of bags can be used, most breeders prefer those that are just wide enough to cover the emasculated florets and long enough to secure to the culm below the emasculated florets.
2. A 70 % ethanolic solution is used in a small bottle to sterilize the forceps to avoid pollen contamination.
3. The microelements, vitamins, and hormones can be prepared in a stock solution and added before use.
4. Suitable anthers for pollination will be light yellow and plump and will dehisce pollen within approximately 1 min when removed and placed on the hand.
5. Good results with wide-crossing can be obtained by starting pollination earlier than the time of peak of anthesis.
6. The use of a wet glassine bag or shading following hand pollination improved seed set, particularly under low humidity and high temperature conditions.
7. Temperatures of 25–30 °C during the day and 20–25 °C at night produce excellent crossing results.
8. Medium in a 40 ml glass-tube with a screw cap is convenient in transferring immature embryos.
9. It is also effective when applied to germinating seeds, young seedlings, roots, and to developing meristems.
10. Dissolve 1 g basic fuchsin in 200 ml boiling distilled water. Shake well and cool to 50 °C, filter, add 30 ml of 1 N HCl to filtrate, and finally add 3 g of $K_2S_2O_3$.
11. Add 45 ml of glacial acetic acid to 55 ml of distilled water. Add 2 g carmine and boil gently for 3 h in a reflux condenser. Filter.
12. A few drops of 1 % FeEDTA or ferric chloride aqueous solution can be added to the fixative for better results when the chromosomes are stained with aceto-carmine.
13. Add 15 ml distilled water to 4 mg certified carmine and 1 ml concentrated HCl. Mix and boil gently for 10 min with stirring. Cool. Add 95 ml 85 % ethanol. Filter.

References

1. Ogbonnaya FC, Abdalla O, Mujeeb-Kazi A et al (2013) Synthetic hexaploids: harnessing species of the primary gene pool for wheat improvement. Plant Breed Rev 37:35–122
2. Thomas H, Lawes DA (1968) Evaluation of the use of species hybrids and synthetic amphiploids in the improvement of the oat crop. Euphytica 17:404–413
3. Dilkova M, Jellen EN, Forsberg RA (2000) C-banded karyotypes and meiotic abnormalities in germplasm derived from interploidy crosses in *Avena*. Euphytica 111:175–184

4. Ueno M, Morikawa T (2007) Production of synthetic polyploid oats and detection of C genome rearrangements by GISH and FISH. Breed Sci 57:339–343
5. Rajhathy T, Thomas H (1974) Cytogenetics of oats (*Avena* L). The Genetic Society of Canada, Ottawa, pp 1–90
6. Leggett MJ (1992) Classification and speciation in *Avena*. In: Marshall HG, Scorrells ME (eds) Oat science and technology. ASA and CSSA, Madison, WI, pp 32–51
7. Ladizinsky G (1998) Evolution of selected crop plants, Oats, *Avena* spp (Gramineae-Avenae). In: Plant evolution under domestication. Kluwer Academic, pp 184–188
8. Loskutov IG (2008) On evolutionary pathways of *Avena* species. Genet Resour Crop Evol 55:211–220
9. Acquaah G (2007) Tools in plant breeding, sexual hybridization and wide crosses in plant breeding (Section 5, Chapter 10). In: Principles of plant genetics and breeding. Blackwell Publishing, Hoboken, NJ, pp 165–167
10. Brown CM, Patterson FL (1992) Population development by hybridization, Artificial hybridization procedures. In: Marshall HG, Scorrells ME (eds) Oat science and technology. ASA and CSSA, Madison, WI, pp 618–623
11. Sleper DA, Poehlman JM (2006) Variations in chromosome number (Chapter 5). In: Breeding field crops, 5th edn. Blackwell Publishing, Hoboken, NJ, pp 74–76
12. Darlington CD, La Cour LF (1976) Smears and squashes, Acetic squash method. (Chapter 5, C). In: The handling of chromosomes, 6th edn. George Allen & Uniwin, London, pp 36–39
13. Snow R (1963) Alcoholic hydrochloric acid-carmine as a stain for chromosomes in squash preparations. Stain Technol 38:9–13
14. Marshall HG, Myers WM (1961) A cytogenetic study of certain interspecific *Avena* hybrids and the inheritance of resistance to diploid and tetraploid varieties to races of crown rust. Crop Sci 1:29–34
15. Thomas H, Leggett JM (1974) Chromosome relationships between *Avena sativa* and the two diploid species *A. canariensis* and *A. prostrata*. Can J Genet Cytol 16:889–894
16. Thomas H, Al-Ansari N (1988) Genotypic control of chromosome pairing in amphiploids involving the cultivated oat *Avena sativa* L. Euphytica 37:37–45
17. Thomas H, Jones ML (1964) Cytological studies of pentaploid hybrids and synthetic decaploid in *Avena*. Chromosoma 15:132–139
18. Ladizinsky G (1995) Domestication via hybridization of the wild tetraploid oats *Avena magna* and *A. murphi*. Theor Appl Genet 91:639–646
19. Ladizinsky G (2000) A synthetic hexaploid (2n = 42) oat from the cross of *Avena strigosa* (2n = 14) and domesticated *A. magna* (2n = 28). Euphytica 116:231–235
20. Rines HW, Porter HL, Cralson ML, Ochocki GE (2007) Introgression of crown rust resistance from diploid oat *Avena strigosa* into hexaploid cultivated oat *A. sativa* by two synthetic hexaploid. Euphytica 158:67–79
21. Morikawa T (1995) Transfer of mildew resistance from the wild oat *Avena prostrata* into the cultivated oat. Bull Univ Osaka Prefect Ser B 47:1–10

Part II

In Vitro Cultures and Genetic Transformation

Chapter 5

Manipulation of Oat Protoplasts for Transient Expression Assays

Robyn Roberts, Jincan Zhang, Nicole Mihelich, Danielle Savino, and Aurélie M. Rakotondrafara

Abstract

Oat protoplasts are a useful and convenient system to study transient expression using whole cells. Nucleic acid can rapidly be introduced into live cells, and, depending on the assay, results can be collected the same day. Compared to plant tissue, oat cell suspension cultures provide a simple, high yielding, and consistent means to isolate protoplasts. Here, we describe how to generate an oat cell suspension culture from callus grown on solidified medium, and how to maintain the oat cells in cell suspension culture for protoplast preparation. Following the culturing procedure, we describe how to isolate oat protoplasts from cell suspension culture by enzymatic digestion of the cell walls and to transiently express nucleic acid (DNA or RNA) into the cells by electroporation.

Key words Oat protoplast, Cell callus, Electroporation, Oat cell suspension culture, Transient expression, Nucleic acid

1 Introduction

Protoplasts, which are plant cells that have had their cell walls removed, comprise the nucleus and the surrounding cytoplasmic material and are encompassed by the cell membrane. Protoplasts are a valuable and advantageous system to study membrane biology, cell physiology, signal transduction, plant breeding, and gene expression [1]. They not only maintain the same physiological responses and cellular activities as cells in an intact plant, but they also provide a rapid means to efficiently introduce nucleic acids into cells. Therefore, controlled measurement of transient gene expression can be completed within a short period of time [1, 2]. Protoplasts may be extracted directly from plant tissues or cell suspension culture [2]. The use of cell suspension culture for protoplast isolation offers several advantages. The cell walls of suspension cultured cells can be easily removed by enzymatic digestion. Suspension cultured cells are derived from the same mother

Sebastian Gasparis (ed.), *Oat: Methods and Protocols*, Methods in Molecular Biology, vol. 1536,
DOI 10.1007/978-1-4939-6682-0_5, © Springer Science+Business Media LLC 2017

culture and are subcultured at regular time intervals. When grown on liquid media, they permit vigorous and homogenous cell growth from week to week, which contributes to the reproducibility and consistency of protoplast preparation and yield. It therefore allows for rapid production of uniformly performing protoplasts in large quantity [2]. Culturing in flasks also takes up less space and requires less maintenance than whole plants [1, 3]. This differs from protoplast isolation from plants where plant-to-plant variation must be taken into consideration.

Here, we describe how to generate an oat cell suspension culture from friable callus grown on solidified medium, and then to maintain a subculture system for oat protoplast isolation. We also describe how to transiently introduce nucleic acid into oat protoplasts by electroporation. Although PEG (polyethylene glycol) is often used alternatively to electroporation to chemically introduce nucleic acid into cells, electroporation requires less nucleic acid and fewer cells and is thus more efficient [4, 5].

The major steps in this procedure include: (1) the generation of an oat cell suspension culture from friable callus plated on solidified medium, (2) the weekly maintenance of the oat cell suspension culture, (3) the enzymatic digestion of oat cells for protoplast isolation, (4) the preparation of oat protoplasts for electroporation with nucleic acid, and (5) cell harvest for transient expression. Excluding the weekly subculturing of the oat cell suspension, the entire process, from the digestion of the cell suspension culture, the isolation of the protoplasts to their electroporation with nucleicacid, may be completed in 24 h for transient expression studies.

2 Materials

2.1 Plant Material

1. Friable oat callus growing on solidified MS plate (*Avena sativa* L.).
2. 7-day-old oat cell suspension culture.

2.2 Media and Solutions

1. Artificial sea water (ASW)/0.6 M mannitol:
   (a) *ASW solution*: 311 mM NaCl (18.18 g/L), 18.8 mM $MgSO_4$ (2.26 g/L), 16.7 mM $MgCl_2$ (3.39 g/L $MgCl_2 \cdot 6H_2O$), 10 mM MES (2.13 g/L 2-(4-morpholino) ethane sulfonic acid), 6.9 mM KCl (0.514 g/L), 6.8 mM $CaCl_2$ (1.0 g/L $CaCl_2 \cdot 2H_2O$), 1.75 mM $NaHCO_3$ (0.148 g/L). Adjust volume to 1 L with ultrapure water and pH to 6.0 with NaOH.
   (b) *0.6 M mannitol*: 109.3 g mannitol (182.2 g/mol) in 1 L of ultra pure water.
   (c) *ASW/0.6 M mannitol (1:1 ratio)*: Add 0.6 M mannitol to ASW solution in a 1:1 ratio (v/v). Autoclave and store at 4 °C up to 1 month, or as separate solutions up to 6 months.

2. Vitamin solution, 100×: 10 g *myo*-inositol, 50 mg nicotinic acid, 50 mg pyridoxine·HCl, 50 mg thiamine·HCl. Adjust volume to 1 L with ultra pure water and filter-sterilize the solution using a 0.2 μm syringe filter. Make 10 mL aliquots (recommended) and store at −20 °C up to 1 year.
3. Murashige and Skoog (MS) medium: 4.33 g MS plant salt mixture (no agar), 10 mL of 100× vitamin solution (*see* recipe), 87.6 mM sucrose (30.0 g/L). Adjust volume to 1 L with ultra pure water and pH to 5.7 with NaOH. Aliquot 40 mL of MS in 125 mL flasks, plug each flask with cotton, and cover them in aluminum foil. Autoclave, and store up to 3 months at 4 °C.
4. Murashige and Skoog (MS) medium plus 0.4 M mannitol: 4.33 g MS plant salt mixture (no agar), 10 mL of 100× vitamin solution (*see* recipe), 87.6 mM sucrose (30.0 g/L), 0.4 M mannitol (72.8 g/L). Adjust volume to 1 L with ultra pure water and pH to 5.7 with NaOH. Autoclave and store up to 2 months at 4 °C.
5. Murashige and Skoog (MS) medium plus agar (yields approximately three plates): 0.433 g MS plant salt mixture (no agar), 1.0 mL of 100× vitamin solution (*see* recipe), 87.6 mM sucrose (3.0 g/L), 0.25 g agar substitute gelling agent (Phytagel, Phytagar or Gelzan). Adjust volume to 100 mL with ultra pure water and pH to 5.7 with NaOH. Autoclave and store up to 1 month at 4 °C. We recommend using fresh plates for callus transfers.
6. Electroporation buffer: 10 mg KH_2PO_4, 57.5 mg $Na_2HPO_4{\cdot}7H_2O$, 3.75 g NaCl, 18.2 g mannitol. Adjust volume to 495 mL with ultra pure water. Adjust pH to 7.2 with NaOH. Make 50 mL aliquots (recommended) and autoclave. Store up to 1 month at 4 °C. Prior to use, add 0.4 mL of filter-sterilized 400 mM $CaCl_2$ per 50 mL aliquot.
7. Enzyme solution: 0.8 % (w/v) hemicellulase (Sigma), 0.175 % (w/v) cellulase (Onozuka RS, Yakult Pharmaceuticals), 0.1 % (w/v) driselase (Sigma). Resuspend in ASW/0.6 M mannitol by stirring the solution very slowly (at a rate of approximately 1 rotation per 4 s) at room temperature for 30–45 min to solubilize enzymes. Adjust pH to 5.6–5.7 with NaOH. Filter-sterilize enzyme solution using a 0.2 μm syringe filter. Always prepare fresh and use 40 mL of enzyme per 8 mL of packed cells for a 5:1 ratio of enzyme solution to packed cells.

2.3 Equipment and Other Materials

1. Electroporation device (e.g., Bio-Rad GenePulser Xcell).
2. 4-mm gap size electroporation cuvettes.
3. Sterile petri plates (100 × 15 mm).
4. 6-well sterile culture plates.
5. Incubator/shaker fitted for 125 mL flasks.
6. Swinging bucket centrifuge with inserts for 15 mL/50 mL conical tubes.

7. Light microscope.
8. Hemocytometer.
9. Sterile 0.2 μm syringe filters.

3 Methods

To prevent the contamination of samples, *all steps in this protocol must be performed under sterile conditions* (laminar flow hood) unless otherwise noted. Media can be easily contaminated without precaution.

3.1 Generation of Oat Suspension Culture from Callus and Its Maintenance

Perform all steps at room temperature unless otherwise noted.

3.1.1 Starting a Liquid Oat Cell Suspension Culture from Friable Callus

1. Identify actively growing oat callus grown on solidified MS medium (maintained at room temperature and excluded from light) (Fig. 1a) (*see* **Note 1**).
2. Use a sterile spatula to gently scrape the friable cell aggregate off of at least half of the entire plate (*see* **Note 2**).
3. Place the clumped cells into 40 mL of fresh MS medium in a 125 mL flask plugged with sterile cotton and covered with aluminum foil.
4. Incubate the cells in the dark at 20–23 °C with flask rotation at about 200 rpm.
5. Proceed with weekly subculturing as described in Subheading 3.1.2 (*see* **Note 3**).

Fig. 1 Oat cells maintained on solid plate and in liquid medium. (**a**) 4-week-old oat friable and actively growing callus on solidified MS medium. (**b**) A 7-day-old actively growing oat cell suspension culture in a 125 mL flask containing MS medium

3.1.2 Maintenance of the Oat Cell Suspension Culture with Weekly Subculture

1. Using a 10 mL serological pipette, remove 10 mL of a 50 mL 7-day-old oat suspension culture (Fig. 1b) while gently swirling the flask to prevent cell settling (*see* **Note 4**).
2. Transfer the 10 mL subculture of cells into 40 mL of fresh MS medium in a 125 mL Erlenmeyer flask.
3. Plug the mouth of the flask with sterile cotton and cover with aluminum foil.
4. Incubate in the dark at 20–23 °C with flask rotation of about 200 rpm.
5. Subculture cells every 7 days (*see* **Note 5**).
6. The remaining cell culture can be next used for digestion, making callus, or discarded.

3.1.3 Maintenance of Viable Callus on Solidified Medium

1. For long-term storage of the callus, select a viable and growing callus plate that is free of contamination (*see* **Note 1**).
2. Use a sterilized spatula to gently scrape oat cell clumps from the old callus plate.
3. Gently move the cell clumps to a fresh MS plate. Spread cells and leave space for future growth. Control pressure applied when spreading to avoid crushing the cells or the medium. At least three plates should be prepared to ensure callus survival.
4. Seal plates with Parafilm and cover with aluminum foil to prevent light exposure.
5. Store the plates in the dark at room temperature. New growth should be visible within 1 week.
6. Repeat the transfer into fresh MS plus agar plate every 4–6 weeks. Any contaminated plates shall be discarded.

3.1.4 Generation of Callus on Solidified Medium from Oat Cell Suspension Culture

1. Remove a 7-day old oat cell suspension culture from the incubator and let the cells settle at the bottom of the flask (approximately 5 min).
2. Gently remove the supernatant by aspiration.
3. Use a sterilized spatula to lift some cells from the bottom of the flask.
4. Gently move the cells onto a fresh MS plate. Spread the cells to leave space for future growth. Control pressure applied when spreading to avoid crushing the cells. At least three plates should be prepared to ensure callus survival.
5. Seal the plates with Parafilm and cover with aluminum foil to prevent light exposure.
6. Store the plates face down in the dark at room temperature. New growth should be visible within 1 week. Actively growing callus can be used to make new MS plate or suspension culture when needed.

3.2 Preparation of Oat Protoplasts

3.2.1 Digestion of the Cell Wall

1. Slowly transfer a 7-day-old oat cell suspension culture into a sterile 50 mL conical tube using a 50 mL serological pipette. Pipette suspension culture slowly to avoid cell breakage.
2. Place the conical tube in a storage rack and let the cells settle for approximately 5 min.
3. Estimate the volume of the packed cells to calculate the volume of enzyme solution needed for digestion for a final 5:1 ratio of enzyme solution to packed cells. Typically, one flask of actively growing cells yields up to 7 mL of packed cells prior to the removal of 10 mL of cell suspension used for subculture (*see* **Note 6**).
4. Pipette off the supernatant. Add half of the required enzyme solution to the packed cells by slowly pouring the solution against the sidewall of the tube, and suspend the cells by slowly pipetting (about 3 s per 1 mL in a 10 mL serological pipette), or by inverting the tube (about one inversion per 2 s).
5. Aliquot the cell solution into at least three sterile standard-sized petri dishes. The final volume in each dish should not exceed 15 mL. Use the remaining enzyme solution to rinse out the conical tube, and aliquot the rinse solution equally amongst the plates (*see* **Note 7**).
6. Seal the plates with Parafilm and wrap them in aluminum foil to prevent light exposure.
7. Using a gyrotory shaker, mix the cell solution at room temperature overnight (16–18 h) at 42 rpm.
8. After 16–18 h, view the digested cells under a microscope to ensure that they are fully digested (Fig. 2a). If cells do not digest completely, do not proceed with the the next step (*see* **Note 8**).

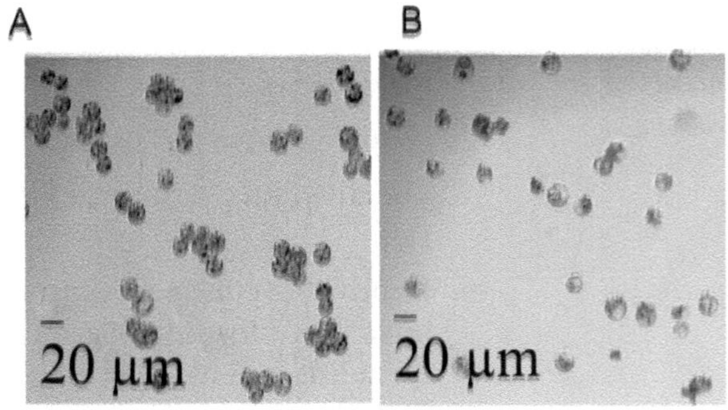

Fig. 2 Oat protoplasts. (**a**) Digested oat cells after 16 h of incubation with the enzyme solution, viewed under a light microscope. Note that protoplasts are spherical once the cell walls are removed. Some partially digested cell clumps can still be observed. However, they can easily be dissociated following washes and pipetting. (**b**) Oat protoplasts immediately after electroporation in MS + 0.4 M mannitol media. Note the mixture of both live (spherical/intact) and dead cells (broken)

3.2.2 Washing Protoplasts

During the washing procedure, cells, washing buffer, and conical tubes should be kept on ice at all times. Prechill centrifuge to 4 °C before beginning experiment (*see* **Note 9**).

1. Collect the digested cells with a sterile 10 mL serological pipette and slowly dispense the cells into two sterile 50 mL conical tubes. If starting with 4 mL packed cells or less, one 50 mL conical tube may suffice. Make sure tubes have an equal volume of protoplast suspension.
2. Centrifuge digested cells at 100 × g for 5 min at 4 °C.
3. Remove supernatant by aspiration using a 10 mL serological pipette, while being careful to not disturb or remove the pelleted cells.
4. Gently resuspend the digested cells in 10 mL of ASW/0.6 M mannitol by pipetting slowly (about 3 s per 1 mL of cell mixture) or by inverting the tube (about one inversion per 5 s).
5. Centrifuge cells at 100 × g for 5 min at 4 °C.
6. Remove the supernatant using a 10 mL serological pipette, and resuspend cells again in 10 mL of ASW/0.6 M mannitol by pipetting (about 3 s per 1 mL of cell mixture) or tube inversion (approximately 1 inversion per 5 s).
7. View an aliquot of the cell suspension under the microscope to ensure that the protoplasts are intact and properly washed with no visible debris left by before proceeding with the next step (*see* **Note 10**).
8. If the protoplasts look good, perform a final centrifugation of the cells at 100 × g for 5 min at 4 °C for preparation to electroporation.

3.3 Electroporation of Oat Protoplasts

3.3.1 Preparation of Cells for Electroporation (See Note 11)

1. Prepare the electroporation solution by adding 0.2 mM spermidine to the electroporation buffer (make sure that the electroporation buffer has the 4 mM $CaCl_2$ added). Add 2 μL of 1 M spermidine stock per 10 mL of electroporation buffer. Place the electroporation solution on ice.
2. Next, slowly remove the supernatant from the final wash of the cells.
3. Gently resuspend the cell pellet in 10 mL of the electroporation solution by slowly pipetting or inverting the tube.
4. Take a cell aliquot and calculate the cell density of the protoplast suspension using a hemocytometer (*see* **Note 12**).
5. Centrifuge the protoplasts at 100 × g for 5 min at 4 °C.
6. Resuspend the cell pellet with the calculated volume of electroporation solution needed for a final cell density of ~2.5×10^6 cells/mL. Place the protoplast suspension on ice.

7. After resuspension, inspect a drop of the cells on a slide using a light microscope to ensure that the cells remain intact (*see* **Note 13**).

3.3.2 Electroporation of Protoplasts with RNA or DNA Samples

During electroporation, all materials should be kept on ice. It is recommended that these steps be performed with no major time delays to increase the chances of cell survival (*see* **Note 14**).

1. In a 6-well culture plate, add 5 mL of MS medium + 0.4 M mannitol to each of the appropriate wells and label one well for each corresponding sample. Set the plate aside at room temperature.
2. Resuspend protoplasts by slowly pipetting or inverting the tube.
3. Use a 1 mL wide-bore pipette tip to aliquot 1 mL of protoplast suspension into a prechilled 4-mm electroporation cuvette. Keep the cuvettes on ice.
4. Proceed with the nucleic acid samples to be tested, one at a time.
5. Suspend the desired amount of DNA or RNA to use for each sample in at least 20 μL of nuclease-free water for easier resuspension.
6. Add the first nucleic acid sample to the aliquoted protoplasts. Mix by gently inverting or tapping the cuvette and follow immediately with electroporation. If using the Bio-Rad GenePulser Xcell, the optimal setting is 300 V for 5 ms on the "square wave" function.
7. After electroporation, use a 1 mL pipette to aliquot 1 mL of MS + 0.4 M mannitol from the 6-well culture plate. Slowly pour the electroporated cells into the sample well, and use the 1 mL medium aliquot to rinse the inside of the cuvette (*see* **Note 15**).
8. Proceed with the remaining samples.
9. After all samples are electroporated, inspect the electroporated cells in the 6-well plate under a light microscope. Approximately 70% of the protoplasts should still be intact after electroporation (Fig. 2b).
10. Use Parafilm to seal the 6-well culture plates, and cover the plates with foil to eliminate light exposure. Leave the plates at room temperature until collection (*see* **Note 16**).

3.4 Collection of Oat Protoplasts (*See* Note 17)

1. Following the required incubation time, inspect the electroporated samples in the respective culture plates using a light microscope to check for cell viability (*see* **Note 18**).
2. Collect cells from the culture plate using a 10 mL pipette and transfer each sample into a separate prechilled 15 mL conical tube. Keep samples on ice.

3. Pellet cells by centrifugation at 100 × *g* for 5 min at 4 °C.
4. Place samples on ice and remove the supernatant using a 10 mL pipette.
5. Proceed with the specific assay (*see* **Note 19** for transient expression of luciferase reporters).

4 Notes

1. An actively growing callus should appear lightly yellow; an orange tint indicates cell death. Do not use callus older than 2 months. If the callus plates are contaminated or if the cells are nonviable, proceed instead with Subheading 3.1.4. Friable callus can be obtained by contacting the authors.
2. At least half of a plate and up to an entire plate of the viable cell callus should be used, depending on the age and growth rate of the cells.
3. It may take up to 4 consecutive weeks of weekly subculture to achieve a densely growing culture that can be used for digestion, depending on the amount and quality of callus used to start the liquid culture. A dense, slightly yellow solution with small visible cell clumps (clumps that are < 1 mm in diameter) is a positive indicator of growth (Fig. 1b). Large, white particles indicate poor cell growth.

   We recommend subculturing at least two flasks per week, in case one becomes contaminated. To produce enough cells for downstream assays, the number of flasks needed per week should be calculated before subculturing.
4. To avoid large cell clumps (and therefore improve cell wall digestion efficiency), a 10 mL serological pipette is recommended for subculturing, which has a narrower bore compared to the 50 mL serological pipette. For particularly clumpy cultures, gently pipette the suspension up and down to break clumps or use progressively narrower serological pipette bores from week to week.
5. Maintenance of viable cell suspension cultures is crucial for generating uniform batches of oat protoplasts. Cells should be subcultured from liquid culture every 7 days into fresh MS medium at a 1:4 ratio of cell culture and medium. Subculturing can be performed up to 9 days after the previous subculture to maintain a live culture. However, we recommend not using cell suspension culture older than 7 days for protoplast isolation. Older cells may resist cell wall digestion and thus result in poor protoplast isolation. A properly maintained subculture should appear slightly yellow, largely devoid of big clumps, and should provide a consistent number of cells from week to week (Fig. 1b).

6. Properly prepare the enzyme solution. The enzymes used in the enzyme solution are sensitive to salt concentrations, osmotic conditions, and pH. Mix solution using a stir bar, at a rate of approximately one rotation per 4 s. It is recommended to closely follow the supplier source indicated for each enzyme, as the authors have found significant differences in the activities of the enzymes from different suppliers.
7. Dividing the cell suspension into several petri dishes allows for a larger surface area to volume ratio for improved digestion. For 3–7 mL of packed cells, three plates are sufficient. The number of plates may need to be adjusted based on the volume of packed cells.
8. After digestion, the cell culture should have a yellowish homozygous cloudy look with few visible clumps with the naked eye. When viewed under a light microscope, cells should no longer appear in large clumps, and intact individual protoplasts should be spherical in shape (*see* Fig. 2a). You may observe partially digested cell clumps. Such cell clumps may be easily dissociated with the downstream washes and pipetting. If most cells are incompletely digested, discontinue the procedure. *See* the Troubleshooting Table for tips on improving the cell wall digestion.
9. After digestion of the cell wall, oat protoplasts are quite fragile and sensitive to osmotic differences and physical damage. Therefore, it is imperative that cells are handled gently and that buffers are prepared with care. Cell breakage reduces the viable cell density, increases the chance of degradation of the nucleic acid sample upon addition to the cells, and may trigger a cascade of death of the surrounding cells. After digestion, the protocol should be completed in a timely manner without any major time breaks, with limited transportation/shuttling of cells to prevent breakage by jarring and with care handling of the cells. Use wide-bore micropipette tips and serological pipettes for cell pipetting. If wide-bore micropipette tips are unavailable, use sterile scissors to cut off the end of a 1 mL pipette tip to simulate a wide-bore micropipette tip. If at any time during the procedure there is a visible color change in cells (from yellow to white), the cells have died and the protocol should be discontinued. *See* the Troubleshooting Table for tips.
10. After centrifugation, live protoplasts should have a yellow tint of color in the pellet. A white pellet indicates that the protoplasts are dead; if observed, discontinue the assay. Following the series of washes, little to no cell debris should be observed under the microscope. The number of protoplasts may be reduced. If some clumps of partially digested cells remain, use a 10 mL serological pipette to dissociate clumps. This should be done by slowly pipetting protoplasts against the sidewall of the conical tube. If unable to disrupt the large clumps, discontinue with the procedure.

Troubleshooting table

Troubleshooting table describing possible problems and solutions for the isolation and electroporation of oat protoplasts

Problems	Possible explanations	Solutions
Generation of oat suspension culture from callus and its maintenance		
Callus isn't growing on the solidified MS medium	1. the MS medium is old or contaminated 2. the callus used for transfer was too old or contaminated	1. properly prepare new MS medium 2. ensure that the starter callus is no more than 4 to 8 weeks old, yellow in color, and void of contamination. If needed, start new callus from liquid culture
Callus transfer to suspension culture isn't growing or is growing slowly	1. the MS medium is old or contaminated 2. the callus used for transfer was too old or contaminated 3. too few cells were used to start the suspension culture	1. properly prepare new MS medium 2. ensure that the starter callus is no more than 4 to 8 weeks old, yellow in color, and void of contamination. If needed, start new callus from liquid culture 3. increase the number of cells lifted from the callus plate to start the suspension culture
Cell suspension culture isn't growing, or isn't growing to proper density	1. improper incubator settings (temperature, rotation) 2. improperly made MS medium 3. MS medium is old or contaminated 4. subculture was performed too early or too late 5. starter culture is dead	1. check incubator settings (20° to 23 °C, 160 to 220 rpm) 2-3. properly prepare new MS medium 4. subculture every 7 days 5. start a new liquid suspension culture from callus
Cells in suspension culture are clumpy; the clumps are more than 1 mm in diameter.	1. improper incubator settings (temperature, rotation) 2. improperly made MS medium 3. MS medium is old or contaminated 4. subculture was performed too late or inappropriately	1. check incubator settings (20° to 23 °C, 160 to 220 rpm) 2-3. properly prepare new MS medium 4. subculture every 7 days; use a smaller bore serological pipette (10 mL or less) to select for fewer and smaller clumped cells during subculture
Cells have an orange tint rather than being light yellow in suspension.	1. improper incubator settings (temperature, rotation) 2. improperly made MS medium 3. medium is old or contaminated 4. subculture is performed too late 5. the cells are dead	1. check incubator settings (20° to 23 °C, 160 to 220 rpm) 2-3. properly prepare new MS medium 4. subculture every 7 days 5. start a new liquid suspension culture from callus
Cells grown on MS medium or in suspension culture are green	light is leaching into cultures/plates	ensure that the cell suspension cultures are grown in the dark, and/or the callus plates are wrapped in foil

(continued)

(continued)

Problems	Possible explanations	Solutions
Preparation of oat protoplasts		
Incomplete digestion of cell wall	1. cell suspension culture is too old 2. improperly made enzyme solution 3. improper overnight rotation during cell wall digestion 4. digestion time is too short 5. cells are too clumpy	1. only use 7-day-old suspension cultures for cell digestion 2. always use freshly prepared enzyme solution. Check the pH of the enzyme solution, and mix the enzyme solution slowly during preparation 3. rotate cells in the enzyme solution at 42 rpm overnight in the dark 4. digest cells 16 to 18 h 5. avoid clumped cells in digestion; use a smaller bore serological pipette (10 mL or less) to perform subculture and select for fewer clumped cells that are less than 1 mm in diameter
Electroporation of oat protoplasts		
Consistent cell death and/or debris during washes	1. improperly made ASW/0.6 M mannitol medium 2. harsh cell resuspension 3. harsh centrifugation 4. cells were over-digested 5. cells not kept on ice	1. properly remake the ASW/ 0.6 M mannitol medium 2. repeat the wash with the ASW/0.6 M mannitol by gently pipetting or slowly inverting the conical tubes to resuspend cells 3. centrifuge at $100 \times g$ at 4 °C 4. digest cells no more than 16 to 18 h 5. keep cells on ice at all times

High proportion of dead cells and/or cell debris over intact cells following electroporation	1. improperly made electroporation buffer medium 2. improper electroporation settings 3. cells have been sitting in the electroporation buffer too long 4. harsh handling of cells immediately before or after electroporation 5. cuvettes and/or cells not kept on ice 6. cells not placed in MS medium directly after electroporation	1. properly remake the electroporation buffer 2. check electroporation settings (300 V, 5 ms for the Bio-Rad GenePulser Xcell) and optimize as needed for differences in machines 3. proceed through all steps with minimal delays 4. gently invert or tap the cuvettes to resuspend cells and the nucleic acid sample immediately before electroporation. After electroporation, gently pour the cuvette contents into the 6-well culture plates and use 1 mL of MS medium to rinse the cuvette 5. keep cuvettes and/or cells on ice. Prechill the cuvettes 6. transfer electroporated cells to MS medium immediately after electroporation
High ratio of dead cells and/or cell debris over intact cells after incubation (before harvest)	improperly made MS+ 0.4 M mannitol medium	properly remake medium and ensure that there is no contamination
No transient expression read-out	1. incomplete cell wall digestion 2. unsuccessful electroporation due to improper electroporator settings 3. poor quality of nucleic acid sample 4. degradation of nucleic acid sample	1. make sure cells are fully digested before proceeding with experimentation. Use 7-day-old suspension culture 2. check electroporation settings (300 V, 5 ms for the Bio-Rad GenePulser Xcell) and optimize as needed for differences in machines 3. check nucleic acid quality before electroporation 4. proceed immediately to electroporation after mixing the nucleic acid with the protoplasts

11. The cells are very sensitive to minute changes in salt concentrations, osmotic conditions, and pH of the electroporation buffer. Inappropriately prepared media may result in cell death upon suspension in the electroporation solution. Use only ultra pure water in the buffer and prepare carefully. The electroporation buffer can become easily contaminated, so it is recommended that small 50–100 mL aliquots of the buffer are made to reduce the risk of contamination. Only open the buffer under sterile conditions (laminar flow hood), and ensure that the buffer is free from contaminants before use.
12. A 10 μL sample should suffice for cell counting, though a 1:10 or 1:100 dilution of cells in electroporation buffer may be required depending on the cell density. Calculate the cell density as follows:

$$\text{measeured cell density} = \frac{\text{average number of cells per small square} \cdot \text{dilution factor}}{\text{volume of the small square}}$$

Then, use the following equation to calculate how much more medium is required to reach ~2.5×10^6 cells/mL, which is sufficient for transient expression. For assays that require longer incubation time such as viral replication studies, scale up the concentration to ~6×10^6 cells/mL. Since 10 mL of electroporation buffer was initially added to the cells for counting, this will need to be subtracted from the final volume.

$$\text{final volume} = \frac{\text{measured cell density}}{2.5 \times 10^6 \text{ cells / mL}} \cdot \text{initial volume}$$

Count the cells in the selected squares until the total count is approximately 100 cells, which is the minimum number of cells needed for a statistically significant count.
13. If there is still a significant amount of debris, perform an additional wash in electroporation buffer + 0.2 mM spermidine before proceeding. If debris is still observed after the additional wash, discontinue the assay. *See* the Troubleshooting Table for tips.
14. Perform steps in a laminar flow hood whenever possible. For transient expression, it is not critical to perform the following steps under sterile conditions. When pipetting cells, use a wide-bore pipette tip to reduce cell damage. Pipette slowly and carefully. Place the 4-mm gap size electroporation cuvettes on ice for at least 5 min before adding cells to cuvettes. Cells should be electroporated immediately upon the addition of the nucleic acid to avoid nucleic acid sample degradation. Handle each sample one at a time to prevent RNA degradation. Use at least 2 pmol of DNA or in vitro transcribed RNA for transient expression

studies. If co-electroporating more than one RNA sample in a single cuvette, mix both RNAs together prior to addition to the protoplasts. For viral replication studies, as much as 20 pmol of nucleic acid may be required. This protocol uses the Bio-Rad GenePulser Xcell for electroporation; however, settings may need to be optimized depending on the device used.

15. A properly executed electroporation should yield a sample with a "stringy" white appearance within the cell sample, which indicates some cell death.
16. Handle the cells with care; protoplasts are fragile after electroporation. For transient expression, cells may be harvested 3–4 h post-electroporation. For viral replication studies, cells may be harvested after 24–48 h for detection. However, some optimization may be required depending on the experimental application.
17. Collection of protoplasts does not need to be performed under sterile conditions, unless required for downstream application. It is recommended that downstream procedures be performed immediately after collection of protoplasts.
18. For shorter incubation times (3–4 h), cell viability should be similar to what was observed after electroporation. For longer incubation times (24–48 h), approximately 50% or more of the cells should be viable.
19. The authors regularly use firefly luciferase reporter RNA to study gene expression. The test RNA is co-electroporated with a Renilla luciferase reporter RNA at a 10:1 ratio for normalization. The following is an example of how the authors proceed with measuring luciferase activity.

Harvesting protoplasts for transient expression of RNA with luciferase reporters

(a) After protoplast collection, add 500 μL of 2× passive lysis buffer (Promega) to each sample in the 15 mL conical tubes.

(b) Resuspend the pellet by inverting the tube or by pipetting with a 1 mL-wide bore pipette. Transfer the suspensions to 2.0 mL microcentrifuge tubes by pipetting.

(c) Pipette cell suspension up and down to break cells.

(d) Shake samples at 2000 rpm on a thermomixer or shaker at 4 °C for 15 min to ensure cell breakage.

(e) Centrifuge the cells at 100 × *g* at 4 °C for 5 min.

(f) Use 100 μL of the supernatant to measure luciferase activity using a luminometer. The authors use a Berthold Technologies Centro XS^3 luminometer and inject 50 μL of the luciferase assay substrate (firefly) followed by 50 μL of Stop&Glo reagent (Renilla) (Promega) to each 100 μL sample of protoplast cell lysate.

References

1. Davey MR, Anthony P, Power JB, Lowe KC (2005) Plant protoplasts: status and biotechnological perspectives. Biotechnol Adv 23(2):131–171. doi:10.1016/j.biotechadv.2004.09.008
2. Rakotondrafara AM, Jackson JR, Kneller EP, Miller WA (2007) Preparation and electroporation of oat protoplasts from cell suspension culture. Curr Protoc Microbiol Chapter 16:Unit 16D.13. doi:10.1002/9780471729259.mc16d03s05
3. Bates GW (1999) Plant transformation via protoplast electroporation. Methods Mol Biol 111:359–366. doi:10.1385/1-59259-583-9:359
4. Vain P, De Buyser J, Bui Trang V, Haicour R, Henry Y (1995) Foreign gene delivery into monocotyledonous species. Biotechnol Adv 13(4):653–671
5. Miao Y, Jiang L (2007) Transient expression of fluorescent fusion proteins in protoplasts of suspension cultured cells. Nat Protoc 2(10):2348–2353. doi:10.1038/nprot.2007.360

Chapter 6

Oat Anther Culture and Use of DH-Lines for Genetic Mapping

Elina Kiviharju, Sirpa Moisander, and Pirjo Tanhuanpää

Abstract

Possibility to make doubled haploids (DHs) from varying crossing populations is a useful tool for enhancing cultivar breeding, and a source of valuable material for genetic research. Oats is reported to be recalcitrant in anther culture with low response and genotype dependency. However, the best recoveries reported have reached up to 30 green regenerants per 100 isolated anthers, which clearly addresses the potential of this technique. In this chapter, one successful oat anther culture protocol is described in detail. Due to the total homozygosity reached in one generation, DH-lines are also an excellent material for genetic mapping. In this chapter, the use of DH-mapping population for marker analyses and linkage mapping is presented.

Key words Oats, *Avena sativa*, Anther culture, Microspore, Regeneration, DH, Haploid, DNA markers, Linkage mapping

1 Introduction

Doubled haploid (DH) technologies are widely used to enhance cultivar breeding programs [1]. Common ways to produce DH-lines are via anther or microspores culture, or through wide crosses. All these approaches have been reported in oats [2–17]. DHs are also desired material for genetic and genomic research and development of DNA markers for marker assisted selection. They represent a permanent population type with the possibility of trait measurements in several locations and in several years. In addition, due to the homozygosity of all individuals, there is no difference in information content in mapping between dominant and codominant markers.

In this chapter, one anther culture protocol [18–20] for producing oat DH-lines is described in detail. In addition, use of a DH progeny for constructing the first doubled haploid linkage map in oat [21, 22] with the protocols for all used marker types is described (*see* Subheading 3.2). The linkage map was subsequently used in positioning of QTL (quantitative trait loci) for important breeding traits on the map [22, 23].

Sebastian Gasparis (ed.), *Oat: Methods and Protocols*, Methods in Molecular Biology, vol. 1536,
DOI 10.1007/978-1-4939-6682-0_6,

2 Materials

2.1 Growing the Anther Donor Plants

1. Sow the seeds of the selected genotypes in a peat–soil mix (two seeds per 11 cm Ø pots, or three seeds per 14 cm Ø pot).
2. Place the pots in the greenhouse and grow in rather low temperature, 16–18 °C day and 12–14 °C night, with a 16/8 photoperiod. Supplement light with fluorescent lamps (or equal) with the total intensity around 350 mmol m^2/s (*see* **Note 1**).
3. Use fertilizer having nitrogen, phosphorus, and potassium balances recommended for cereals (*see* **Note 2**).

2.2 Stock Solutions for Anther Culture Media

1. Preparation of the stock solutions for W_{14}-medium [24] is shown in the Table 1. MS-Fe-EDTA is prepared by solving 2.785 g $FeSO_4{\cdot}7H_2O$ and 3.725 g Na_2-EDTA (titriplex) separately in distilled water, combining these solutions, and heating to the boiling point. Note that $FeSO_4{\cdot}7H_2O$ solves completely only after the both solutions are combined.

   All solutions are made for distilled or the same purity level water. Stock solutions are autoclaved for 20 min at 121 °C temperature and 1 kg/cm^2 pressure, and stored in the refrigerator at +4 °C.
2. Preparation of stock solutions for MS-medium [25] are described here.

   MS macrosalts (10×): 19 g/L KNO_3, 16.5 g/L NH_4NO_3, 4.4 g/L $CaCl_2{\cdot}2H_2O$, 3.7 g/L $MgSO_4{\cdot}7H_2O$, and 1.7 g/L KH_2PO_4, and Fe-EDTA as described above.

   MS microsalts (1000×): 2230 mg/L $MnSO_4{\cdot}4\ H_2O$, 620 mg/L H_3BO_3, 1060 mg/L $ZnSO_4{\cdot}7H_2O$ (860 mg/L $ZnSO_4{\cdot}4H_2O$ in the original protocol), 83 mg/L KI, 25 mg/L $Na_2MoO_4{\cdot}2H_2O$, 2.5 mg/L $CuSO_4{\cdot}5H_2O$ (1.6 mg/L $CuSO_4$ in the original protocol), and 2.5 mg/L $CoCl_2{\cdot}6\ H_2O$.

   MS vitamins (100×): 20 mg/100 ml glycine, 1000 mg/100 ml myo-inositol, 5.0 mg/100 ml nicotinic acid, 5.0 mg/100 ml pyridoxine–HCl, and 1.0 mg/100 ml thiamine–HCl.
3. Growth regulators: Stocks solutions of 1 mg/ml are prepared, pipetted to Eppendorf tubes and stored in the freezer at −20 °C. 2,4-D (2,4-dichlorophenoxyacetic acid) is first solved to a small volume of ethanol (95–97%) by warming, before adding distilled water. NAA (α-naphthaleneacetic acid), kinetin, and BAP (6-benzylaminopurine) are dissolved to 1 N KOH before adding water. Ethephon™ [(2-chloroethyl)phosphonic acid] and L-cysteine are dissolved directly to water. Ethephon™ have commercially been produced by Dr. Ehrenstorfer GmbH, Augsburg.
4. For a liquid medium use Ficoll® 400 (Pharmacia Biotech).

Table 1
Preparation of stock solutions for W_{14} medium

Stock solution	Composition	mg/L in medium	Stock solution composition, solved to water
			10× stock solution, volume 1 L
W_{14} macro salts	KNO_3	2000	20 g
	$NH_4H_2PO_4$	380	3.8 g
	$MgSO_4 \times 7H_2O$	200	2.0 g
	$CaCl_2 \times 2H_2O$	140	1.4 g
	K_2SO_4	700	7.0 g
			100× stock solution, volume 250 ml
W_{14} micro salts	$MnSO_4 \times 4H_2O$	8.0	200 mg
	$ZnSO_4 \times 7H_2O$	3.0	75 mg
	H_3BO_3	3.0	75 mg
	KI	0.5	12.5 mg
	$CuSO_4 \times 5H_2O$	0.025	0.625 mg
	$CoCl_2 \times 6H_2O$	0.025	0.625 mg
	$NaMoO_4 \times 2H_2O$	0.005	0.125 mg
			100× stock solution, volume 200 ml
W_{14} vitamins	Glycine	2.0	40 mg
	Thiamine	2.0	40 mg
	Pyridoxine	0.5	10 mg
	Nicotinic acid	0.5	10 mg
MS-Fe-EDTA	$FeSO_4 \times 7H_2O$	27.8	
	$NaEDTA \times 2H_2O$	37.3	

5. Colchicine solution: 0.05 % (w/v) colchicine in 2 % DMSO is prepared. For 15 samples: 0.075 g colchicine, 147 ml water, and 3 ml DMSO are mixed. Use green nitrile gloves and fume hood.

2.3 Media Preparation for Anther Culture

2.3.1 Induction Medium with Solid and Liquid Layers

1. Composition of the oat anther culture induction medium is shown in Table 2 (*see* **Note 3**). Double layer medium, with solid and liquid phases is used. Solid part is solidified with 0.3 % Phytagel™ (Sigma), liquid part contains 10 % Ficoll® 400 (Pharmacia Biotech) (*see* **Note 4**).
2. To make 250 ml of the solid induction medium: Phytagel™ is added to 150 ml of distilled water and autoclaved for 20 min

Table 2
Preparation of 250 ml or 1000 ml of solid and liquid induction medium for oat anther culture

	/1000 ml	/250 ml
Solid part:		
W_{14}-macro salts 10×	100 ml	25 ml
W_{14}-micro salts 100×	10 ml	2.5 ml
W_{14}-vitamins 100×	10 ml	2.5 ml
MS-Fe-EDTA	5 ml	1.25 ml
Maltose × H_2O	105 g	26.25 g
2,4-D	5.0 mg	1.25 mg
BAP or kinetin	0.5 mg	0.125 mg
Ethephon™	20 mg	5.0 mg
L-cysteine	50 mg	12.5 mg
Myo-inositol	500 mg	125.0 mg
Phytagel™	3 g	0.75 g
pH 6.0		
Liquid part:		
W_{14}-macro salts 10×	100 ml	25 ml
W_{14}-micro salts 100×	10 ml	2.5 ml
W_{14}-vitamins 100×	10 ml	2.5 ml
MS-Fe-EDTA	5 ml	1.25 ml
Maltose × H_2O	105 g	26.25 g
2,4-D	5.0 mg	1.25 mg
BAP or Kin	0.5 mg	0.125 mg
Ethephon™	20 mg	5.0 mg
L-cysteine	50 mg	12.5 mg
Myo-inositol	500 mg	125.0 mg
Ficoll® 400	100 g	25.0 g
pH 6.0		

at 121 °C and 1 kg/cm^2. All other components are dissolved into distilled water, volume is fulfilled to 100 ml and pH is adjusted to 6.0. This solution is then mixed to the autoclaved Phytagel™ solution through 0.22 or 0.45 μm syringe filter (*see* **Note 5**).

3. To make 250 ml of the liquid induction medium: All components except Ficoll® are dissolved to distilled water and pH is adjusted to 6.0. Then Ficoll® is added to the solution and the volume is adjusted to 250 ml by distilled water. Solution is heated until Ficoll® is dissolved, and then sterilized by autoclaving. Medium is stored in the refrigerator.
4. For 3.5 ml in diameter petri dishes (Falcon) 2.5 ml of solid medium is added with liquid dispenser, cooled in the laminar hood and stored at the fridge (*see* **Note 6**).
5. Liquid part of 1 ml is pipetted on the top of the solidified medium just before starting to isolate the anthers.

Table 3
Preparation of 250 ml or 1000 ml of regeneration medium for oat anther culture

	/1000 ml	/250 ml
W_{14}-macro salts 10×	100 ml	25 ml
W_{14}-micro salts 100×	10 ml	2.5 ml
W_{14}-vitamins 100×	10 ml	2.5 ml
MS-Fe	5 ml	1.25 ml
NAA	2.0 mg	0.5 mg
Kinetin	0.5 mg	0.125 mg
Sucrose	20 g	5 g
Phytagel™	3 g	0.75 g
pH 6.0		

2.3.2 Regeneration and Rooting Medium

1. Composition of the regeneration medium is shown in Table 3.
2. To make 250 ml of the solid regeneration medium: Phytagel™ is added to 150 ml of distilled water and autoclaved for 20 min at 121 °C and 1 kg/cm^2. All other components are dissolved into distilled water, volume is fulfilled to 100 ml and pH is adjusted to 6.0. This solution is then mixed to the autoclaved Phytagel™ solution through 0.22 or 0.45 μm syringe filter.
3. The rooting medium contains MS-medium salts and vitamins (*see* Subheading 2.2, **item 2**) supplemented with 0.2 mg/L NAA, 2% sucrose and solidified with 0.3% Phytagel. pH is adjusted to 5.8 and medium is sterilized by autoclaving.

2.4 Buffers for Ploidy Level Analysis

1. Buffer-1: 27.2 mg/L KH_2PO_4, 101.1 mg/L KNO_3, 1470.2 mg/L $CaCl_2 \cdot 2H_2O$, 246.6 mg/L $MgSO_4 \cdot 7H_2O$, 0.166 mg/L KI, 0.025 mg/L $CuSO_4 \cdot 5H_2O$, and 145.76 g/L sorbitol. pH is adjusted to 5.6 and solution is autoclaved.
2. Buffer-2: 882 mg/100 ml $C_6H_5Na_3O_7 \cdot 2H_2O$, 915 mg/100 ml $MgCl_2 \cdot 6H_2O$, 462 mg/100 ml MOPS (3-(N-morpholino) propanesulfonic acid). pH is adjusted to 7.0 and solution is autoclaved.
3. Solution-A for 36 samples: 45 ml Buffer-2, 450 ml Triton X-100, 225 μl 2-mercaptoethanol and 900 μl PI (Propidium iodine). Prepare on ice and in a fume hood using gloves.
4. Solution-B for 36 samples: 9 ml Solution-A and 22.5 μl RNAse. Prepared on ice in a fume hood using gloves.
5. Cellulase Onozuka RS 1% solution.

2.5 Buffers for DNA Extraction

1. 2% (w/v) CTAB (cetyltrimethylammonium bromide) buffer pH 8.0: 20 g CTAB, 81.8 g NaCl, 7.44 g EDTA (ethylenediaminetetraacetic acid), and 12.1 g Tris base. Make up to 1000 ml with H_2O. Adjust pH to 8.0. Add 0.2 % (v/v) 2-mercaptoethanol but not until the buffer is being used.
2. TE Buffer: 10 mM Tris–HCl, 1 mM EDTA, pH 8.0.

2.6 Equipment

1. Flow cytometer for ploidy analyses.
2. DNA spectrophotometer for measuring DNA concentrations.
3. PCR thermal cycler for marker amplifications.
4. Capillary sequencer for separating and visualizing markers with fluorescent label.

2.7 Plant Material

A mapping population is created in order to make a linkage map. In our study, we selected a Finnish oat cultivar 'Aslak' and a Swedish cultivar 'Matilda' as crossing parents, since they differed from each other in respect to several traits that we were interested in, such as β-glucan content, oil content, and leaf blotch resistance. Several individuals from parent cultivars were crossed (*see* **Note 7**) with each other to produce two hundred F1 seeds. These seeds were sown for anther donor plants. By using anther culture protocol described above, 148 oat regenerants were produced in three separate anther culture experiments. The final mapping progeny consisted of 137 doubled haploids (*see* **Note 8**). Of these, the genome of 82 individuals (60 %) was doubled by colchicine and 55 were spontaneously doubled haploids.

3 Methods

3.1 Carrying Out the Anther Culture Protocol

Horizontal laminar hood is used for all aseptic work. Distilled water or equal quality water is used.

3.1.1 Collecting Tillers for Cold Treatment

1. Collect oat tillers when the tip of the panicle inside of the leaf sheath can be observed a little bit above the level of the second leaf base (*see* **Note 9**). Usually it takes 5–6 weeks growing time in the greenhouse for the main tillers to reach this growing stage, in the described conditions.
2. Cut tillers under the knot and place them in the Erlenmeyer flasks (or other vessel) containing tap water, cover with plastic bag to prevent evaporation, and place in a cold room for 7 days at +4 °C in dark for cold pretreatment (*see* **Note 10**).

3.1.2 Surface Sterilization of Tillers

1. After cold pretreatment, take tillers to the tissue culture laboratory. Cut stems so that panicles are still enclosed in the leaf sheath (0.5–1 cm over panicle from both sides).

Surface-sterilize stems by wiping with a cloth moistened well with 70% ethanol.

2. Then spilt stem by scalpel, remove panicle and place it onto an autoclaved glass petri dish (or sterile plastic) in the laminar hood. To prevent drying use lid if there is need to wait. Usually no other sterilization is needed for the plant material grown in the greenhouse (*see* **Note 11**).

3.1.3 Determine Developmental Stage of Microspores

1. To check that microspores are in the proper developmental stage, pick some anthers and squash them with glass rod in 0.3 M mannitol (instead aceto-carmine stain or just a drop of water can be used) and look by light microscope. Late uninucleate stage microspores are preferred, due to their assumed better capability to enter the embryogenetic developmental pathway [26] (*see* **Note 12**).
2. Often anthers around 2.5 mm long (varying between 2 and 3 mm) have had the biggest proportion of late uninucleate microspores. Anther length can be easily checked during the isolation process by comparing excised anthers to a piece of millimeter paper, which is placed under a transparent petri dish. Check the microspore developmental stage for each genotype and also for different greenhouse batches of the same genotype, since growing conditions may affect the developmental stage (e.g., weather or watering conditions).

3.1.4 Isolation of Anthers for Induction Medium and Heat Pretreatment

1. One milliliter of liquid induction medium is pipetted onto the top of the solid induction medium just before starting anther isolation.
2. Spikelets are cut from the panicle and opened with scalpel (knife). Anthers are picked by curved sharp forceps (e.g., Aesculab No. BD329R). Handle anthers carefully and do not cut or damage them by pressing too hard.
3. Anthers are placed on the top of the liquid medium layer and they should float (look that they do not drown).
4. Density of 30 anthers per 3.5 cm diameter petri dish (Falcon) is used. Petri dishes should be closed with Parafilm (M). Dishes are placed in a growth cabin for heat pretreatment at +32 °C for 5 days in the dark, before culturing at +28 °C or +25 °C in the dark [5] (*see* **Notes 13** and **14**).

3.1.5 Transfer of the Embryo-like Structures (ELS) or Calli to the Regeneration Medium and Rooting the Developing Plantlets

1. The first embryo-like structures (ELS) are formed usually after 4–5 weeks culture (Fig. 1). They are transferred onto solid regeneration medium after 5–8 weeks in culture, at 1-week intervals at the density of five ELS per 3.5 cm-diameter petri dish. Regeneration cultures are incubated at the same temperature as induction cultures, under low light (approximately 40 μmol/m^2 s) with a 16-h photoperiod (*see* **Note 15**).

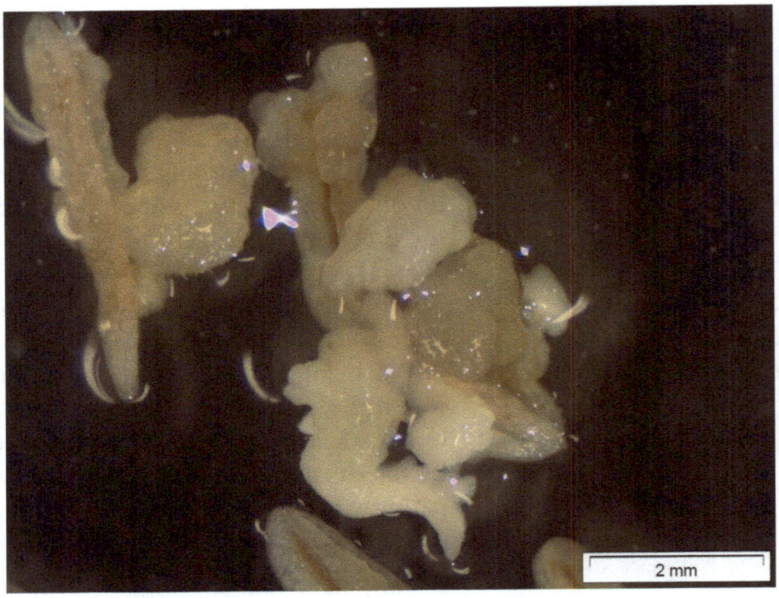

Fig. 1 Anther culture of hulless oat cultivar "Lisbeth" after 6 weeks culture in induction medium. (Photo taken by Elina Kiviharju)

Fig. 2 Plant regeneration in regeneration medium after 1 week of the transfer of ELS, cultivar "Lisbeth." (Photo taken by Elina Kiviharju)

2. Anther response can be assessed by counting how many embryo-like structures are formed by 100 anthers after 8 weeks of culture.
3. Regenerating green plantlets (Fig. 2) are transferred onto solid rooting medium into glass tubes (15 cm long and 2.2 cm wide) or plastic chambers (Magenta™ vessels, 10 cm high, 6.5 cm × 6.5 cm top area) and cultured at 25 °C (or lover) under dim light.

3.1.6 Determination of Ploidy Levels and Colchicine Treatment

1. Ploidy level of the regenerated plants is the most conveniently determined by flow cytometry (*see* **Note 16**). Nuclei are isolated from freshly isolated protoplasts of the leaf tissue, and stained with propidium iodide [27].
2. Add 1–2 ml of Buffer-1 to 3.5 cm petri dishes. Cut two to three leaf pieces of 2 cm long from the regenerated plantlets and place them in petri dishes. Cut leaf species into long narrow stripes with sharp scalpel. Remove buffer and add 1 ml of cellulase solution. Cover petri dishes and put them in a laboratory shaker, 30 rpm, overnight, in the room temperature.
3. Next day, put 41–48 mesh nylon nets into the Eppendorf tubes and filtrate the solution through them by using pipette.
4. Centrifuge tubes at 82 × *g* for 5 min, pipet supernatant carefully and discard it.
5. Add 1 ml Buffer-1 and mix carefully.
6. Centrifuge tubes at 82 × *g* for 3 min, and prepare solution-A.
7. Pipet supernatant and discard it.
8. Add 1 ml of solution-A and mix carefully.
9. Incubate for 10 min, and during that time prepare solution-B.
10. Centrifuge tubes at 16,000 × *g* for 10 s and discard the supernatant.
11. Add 200 μl solution-B and mix carefully. Incubate for 15 min at +37 °C. After that move samples on ice.
12. The DNA content of stained nuclei is analyzed using a laser flow cytometer, e.g., Becton Dickinson FACSort, argon-ion laser at 488 nm wavelength. The barley cultivar Sultan is used as an external standard (4 C = 22.24 pg DNA) [28].
13. DNA content of each sample is determined by comparing position of the nuclear 2 C peak with the peak of an external standard on the histograms.
14. Doubled haploid plants are transferred to the greenhouse (*see* below).

3.1.7 Chromosome Doubling and Transferring Plantlets into the Greenhouse

1. Haploid genomes are treated with colchicine in order to double the chromosomes. First, the solid medium is washed away from the roots with tap water.
2. Roots of the plantlets are cut to be approximately 2–3 cm long and plantlets are placed back in their washed empty glass tubes.
3. Colchicine solution (0.05 % colchicine in 2 % DMSO) is prepared in the fume hood. Add 10 ml colchicine to glass tubes, so that root and meristem are totally covered and under the surface.

4. Tubes are covered with piece of aluminum foil (not each but the whole rack with one piece to prevent evaporation) and treatment continues for 5 h at the room temperature, in fume hood.
5. Then roots are flushed carefully in tap water for five times and finally left under water for 1 h. Use green nitrile gloves through the whole procedure.
6. Plantlets are planted to a peat soil mix to small pots (5–6 cm in diameter) and transferred into the greenhouse, first to acclimatization chambers, propagators, to adapt to the greenhouse conditions. To ensure adequate moisture, water is sprayed when needed. Later plants are moved to 11–12 cm in diameter pots to continue growing. Growing conditions are the same as for the anther-donor plants (*see* **Note 17**).

3.2 Use of DH Population for Linkage Mapping

DNAs are extracted from the leaves of the parents and DH progeny using a CTAB method [29] with a couple of modifications (*see* **Note 18**).

3.2.1 DNA Extraction

1. Grind 1.5–3 g of oat leaves from one plant in liquid nitrogen with a mortar and pestle.
2. Transfer crushed plant material with the aid of a spatula into a bottle containing 15 ml of CTAB buffer prewarmed to 60 °C. Add 30 μl of mercaptoethanol and stir.
3. Incubate the bottle in a water bath with gentle shaking at 60 °C for 30–60 min.
4. Pour the content of the bottle into a centrifuge tube. Add 15 ml of chloroform–isoamyl alcohol (24:1) and invert gently to mix the components.
5. Centrifuge for 10 min at 2200 × *g*.
6. Transfer the upper aqueous phase carefully with a pipette with a wide bore into a new tube.
7. Add 1 μl of 10 mg/ml RNAse per 1 ml DNA solution. Incubate for 30 min at +37 °C.
8. Incubate the tube for 5 min on ice.
9. Precipitate DNA by adding 0.6–1 volume of cold (−20 °C) isopropanol to the tube.
10. Mix by inverting until you can pick the precipitated DNA with a bent glass rod.
11. Transfer DNA to an Eppendorf tube and add 1 ml of 70% ethanol.
12. Centrifuge the tube for 10 min at 10,500 × *g*.
13. Absorb ethanol carefully and wash the pellet once more with 0.5 ml of 70% ethanol.
14. Air-dry the pellet on bench (*see* **Note 19**).

15. Add 150 μl of TE Buffer. Tap the tube gently and let the DNA dissolve overnight (*see* **Note 20**).
16. Measure DNA content with a spectrophotometer.
17. Store DNA at −20 °C.

3.2.2 DNA Markers

DNA markers are first tested in the parents for their functionality and polymorphism. The most polymorphic ones are then analyzed in the DH progeny. DNA markers can be amplified in a reaction volume of 10 μl or 25 μl in 0.2 ml PCR reaction tubes, strips or a 96-well plate (*see* **Note 21**). Amplification products are separated and visualized depending on the marker method either by running on an agarose gel, usually overnight (RAPDs, ISSRs, REMAPs, and IRAPs) or with capillary electrophoresis on an automated DNA sequencer (microsatellites, AFLPs, SRAPs, SNPs). When visualized on a sequencer, one primer of each primer pair is labeled with a fluorescent dye, FAM (5-carboxyfluorescein), HEX (hexachloro-6-carboxyfluorescein), or TET (6-carboxytetrachlorofluorescein).

In each PCR protocol, minor differences to those explained might exist depending on the primer (*see* **Note 22**).

1. *Oat microsatellites*: Oat microsatellites from various sources can be used [30–34].

   (a) Set up a 25 μl PCR reaction for microsatellites from [30–33]:

   - 20–40 ng template DNA
   - 400 nM each primer
   - 0.75 U *Taq* polymerase (MBI Fermentas) or 0.4 U Biotools DNA polymerase (B&M Labs, S.A.)
   - Buffer supplied by the respective enzyme manufacturer.
   - 1.5 (for MBI *Taq*) to 2 mM $MgCl_2$ (Biotools)
   - 100 μM each dNTP

   Depending on microsatellite, two different PCR programs (mainly the same as in ref. 30 but with reduced cycles, *see* **Note 23**) are used.

PCR program 1

18 cycles of:	Denaturation: 94 °C, 1 min
	Annealing: touchdown from 64 °C to 55 °C, decreasing progressively by 0.5 °C every cycle, 30 s
	Extension: 72 °C, 1 min
25 additional cycles of:	Denaturation: 94 °C, 1 min
	Annealing: 55 °C, 1 min
	Extension: 72 °C, 1 min
Extension cycle: 72 °C, 10 min	

PCR program 2 is otherwise the same as PCR program 1 but with a touchdown from 67 °C to 58 °C, and the amount of additional cycles is 15 with an annealing temperature of 58 °C.

(b) Set up a 10 μl reaction for microsatellites from [34]:

- 10 ng template DNA
- 500 nM each primer
- 0.25 U FIREPol DNA polymerase I (Solis BioDyne OU)
- Buffer B with 2.5 mM $MgCl_2$ supplied by the enzyme manufacturer
- 100 μM each dNTP

Two different PCR programs are used (3 and 4). *PCR program 3* is mainly the same as PCR program 1 but with a touchdown from 66 °C to 55 °C, and with extension time 30 s.

PCR program 4

Initial denaturation step: 94 °C, 5 min	
35 cycles of:	Denaturation: 94 °C, 1 min
	Annealing: 55 °C, 1 min
	Extension: 72 °C, 1 min
Extension cycle: 72 °C, 10 min	

PCR amplification products are separated and visualized with capillary electrophoresis on an automated DNA sequencer.

2. *Barley and rye microsatellites*: When using microsatellites from other species, e.g., barley [35–37] and rye (38; Lochow-Petkus GmbH), several different PCR conditions might have to be tested in the oat parents (*see* **Note 24**). In our study, only one functioning and polymorphic barley microsatellite (HVM20) was found.

(a) Set up a 25 μl PCR reaction:

- 40 ng template DNA
- 160 nM each primer
- 0.4 U Red Hot DNA polymerase (Advanced Biotechnologies)
- Buffer supplied by the enzyme manufacturer
- 2.5 mM $MgCl_2$
- 100 μM each dNTP

Basically the same PCR program 1 which is used for most oat microsatellites (*see* **step 1a** in Subheading 3.2.2) is used for HVM20, except that touchdown is from 54 °C to 45 °C and the amount of cycles 18 + 30.

PCR amplification products are separated and visualized with capillary electrophoresis on an automated DNA sequencer.

3. *RAPD*: RAPD (random amplified polymorphic DNA) primers consist of random 10mers or 11mers. Only one primer is used in each PCR reaction.

   (a) Set up a 25 μl PCR reaction:
   - 20 ng template DNA
   - 600 nM primer
   - 0.75 U Biotools polymerase (B&M Labs, S.A.)
   - Buffer supplied by the enzyme manufacturer
   - 2 mM $MgCl_2$
   - 100 μM each dNTP

PCR program

Initial denaturation step: 94 °C, 3 min	
35 cycles of:	Denaturation: 94 °C, 30 s
	Annealing: 38 °C, 30 s
	Extension: 72 °C, 1 min
Extension cycle: 72 °C, 5 min	

PCR amplification products are separated and visualized by running on a 1.4% SeaKem agarose gel.

4. *ISSR*: ISSR (inter simple sequence repeat) technique involves amplification of DNA segments between two identical microsatellite repeat regions and a primer contains this microsatellite repeat sequence anchored at the 3′ end [39]. Only one primer is used in each PCR reaction.

   (a) Set up a 20 μl PCR reaction:
   - 20 ng template DNA
   - 200 nM primer
   - 1.0 U Biotools polymerase (S&M Labs, S.A.)
   - Buffer supplied by the enzyme manufacturer, with 2 mM $MgCl_2$
   - 200 μM each dNTP

PCR program

Initial denaturation step: 94 °C, 2 min	
35 cycles of:	Denaturation: 94 °C, 20 s
	Annealing: 56 °C, 20 s
	Extension: 72 °C, 2 min
Extension cycle: 72 °C, 10 min	

PCR amplification products are separated and visualized by running on a 1.6% agarose gel.

5. *Retrotransposon-based markers: IRAP and REMAP*: IRAP (inter-retrotransposon amplified polymorphism) and REMAP (retrotransposon-microsatellite amplified polymorphism) markers are based on retrotransposons [40–42]: IRAP markers detect polymorphisms in amplification products generated between retrotransposons, and REMAPs between retrotransposons and microsatellites. Primers are designed by cloning retrotransposon regions from the oat genome (also from barley, rye and timothy), identifying long terminal repeats (LTRs), and choosing conserved motifs at or near their termini. In IRAP analysis, one retrotransposon-based primer is used in each PCR reaction; in REMAP analysis, also a microsatellite-based primer. The microsatellite-based primers are the same as used in ISSR analyses.

   (a) Set up a 20 μl PCR reaction for IRAPs:

   - 25 ng template DNA
   - 250 nM primer
   - 1.0 U of Biotools polymerase (S&M Labs, S.A.)
   - 10 μl of Failsafe 2× PCR PreMix D buffer (Epicentre, *see* **Note 25**) containing $MgCl_2$ and dNTPs

   PCR program is the same as for ISSRs but with 45 cycles and annealing temperature of 60 °C.

   PCR amplification products are separated and visualized by running on a 1.6% Seakem agarose gel.

   (b) Set up a 20 μl PCR reaction for REMAPs:

   - 50 ng template DNA
   - 500 nM each primer
   - 1.5 U Biotools polymerase (Biotools B&M Labs)
   - 10 μl Failsafe 2× PCR PreMix D buffer (Epicentre) containing $MgCl_2$ and dNTPs

   PCR program is the same as for ISSRs but with 30 cycles and annealing temperature of 58 °C.

   PCR amplification products are separated and visualized by running on a 1.6% agarose gel.

6. *AFLP* (amplified fragment length polymorphism) [43]

   (a) Restriction and ligation are performed simultaneously in a 30 μl total volume containing

   - 0.5 μg DNA
   - 5 U each restriction enzyme *Eco*RI and *Mse*I
   - NEbuffer2

- 100 μg/ml bovine serum albumin
- 0.4 mM ATP
- 33 U (cohesive-end ligation units) T4 DNA ligase
- 2.5 μM *Mse*I adapters and 0.25 μM *Eco*RI adapters. Adapters are prepared by mixing equimolar amounts of both strands, heating the mixture to 95 °C for 5 min, and then cooling it down to room temperature.

The restriction–ligation reaction mixture is incubated for 3 h at 37 °C, 4 h at 21 °C, and 15 min at 70 °C. To ensure that the reactions have succeeded, 10 μl from a couple of samples is run on a 0.8 % agarose gel.

(b) Pre-selective amplifications are performed in 20 μl reactions containing:

- 50 ng = 3 μl of DNA from the restriction–ligation reaction
- 375 nM primers (*Eco*RI + A, C, or G and *Mse*I + A or C)
- 0.25 U Biotools polymerase (B&M Labs, S.A.)
- Biotools buffer with 2 mM $MgCl_2$
- 200 μM each dNTP

PCR program

Initial incubation: 72 °C, 2 min	
28 cycles of:	Denaturation: 94 °C, 30 s
	Annealing: 56 °C, 1 min
	Extension: 72 °C, 1 min

To ensure that the reactions have succeeded, 5 μl from a couple of samples is run on a 0.8 % agarose gel.

(c) Selective amplifications are performed in 20 μl reactions containing:

- 5 μl from a tenfold diluted pre-selective amplification reaction as template DNA
- 250 nM *Eco*RI primers
- 500 nM *Mse*I primers
- 0.25 U Biotools polymerase (S&M Labs, S.A.)
- Biotools buffer with 2 mM $MgCl_2$
- 200 μM each dNTP

PCR program:

9 cycles of:	Denaturation: 94 °C, 30 s
	Annealing: touchdown from 65 °C to 56 °C, reduced by 1 °C per cycle, 30 s
	Extension: 72 °C, 1 min
28 cycles of:	Denaturation: 94 °C, 30 s
	Annealing: 56 °C, 30 s
	Extension: 72 °C, 1 min

The selective *Eco*RI primers are labeled with a fluorescent dye, FAM, HEX, or TET, and AFLP reactions with different dyes are multiplexed for analysis on a capillary sequencer.

7. *SRAP*: SRAP (sequence-related amplified polymorphisms) markers are amplified from open reading frames [44]. Forward primers ('me') preferentially anneal to exonic regions and reverse primers ('em') to intronic regions and promoters. Primer sequences are presented in [45].

   Set up a 25 μl PCR reaction:

   - 20 ng template DNA
   - 400 nM each primer
   - 0.5 U Biotools polymerase (S&M Labs, S.A.)
   - Buffer supplied by the enzyme manufacturer with 2 mM $MgCl_2$
   - 100 μM each dNTP

   PCR program

5 cycles of:	Denaturation: 94 °C, 1 min
	Annealing: 35 °C, 1 min
	Extension: 72 °C, 1 min
40 cycles of:	Denaturation: 94 °C, 1 min
	Annealing: 47 °C, 1 min
	Extension: 72 °C, 1 min

PCR amplification products are separated and visualized with capillary electrophoresis on an automated DNA sequencer.

8. *DArT*: DArT (Diversity Arrays Technology) microarray-based markers are genotyped by Triticarte Pty Ltd (Australia) as described in [46].
9. *SNP markers*

   (a) An SNP (single nucleotide polymorphism) linked to a dwarfing gene [47].

Set up a 25 μl PCR reaction:

20 ng DNA

200 nM each primer: ACATGCCGTGGCGTGGGATA and ACATGCCGTGCCTCAGTCCTG

0.75 U *Taq* polymerase (MBI Fermentas)

Buffer supplied by the enzyme manufacturer, 1.5–2 mM $MgCl_2$

100 μM each dNTP

PCR program

Initial denaturation step: 95 °C, 2 min	
35 cycles of:	Denaturation: 95 °C, 30 s
	Annealing: 64 °C, 30 s
	Extension: 72 °C, 1 min
Extension cycle: 72 °C, 6 min	

The SNP reaction is carried out using the MegaBACE SNuPe Genotyping Kit (Amersham Pharmacia Biotech) with a SNP primer CCATTGATACATCTTCCACT.

(b) Three SNP analyses for ACCase (acetyl-CoA carboxylase) [23]

The existence of two *Accase* loci complicate the development of the SNP assay and therefore, three different SNP analyses are performed and results combined.

Set up a 25 μl PCR reaction:

20–40 ng DNA

200 nM each primer, for SNP analyses 1 and 2: GACCTCCAGAAGGTTGTAGAC and CTGTGTTACCCGCAGAGCTT and for SNP analysis 3: GAGATGAAGTATACACCGTCAC and GACGATGCAGCATAGTGGTA

0.5 U Biotools DNA polymerase (Biotools B&M Labs, S.A.) or 0.75 U *Taq* polymerase (MBI Fermentas, SNP analysis 3)

Buffer supplied by the enzyme manufacturer with 2 mM $MgCl_2$

50 μM each dNTP

PCR program is the same as when amplifying the SNP linked with the dwarfing gene except that the amount of cycles for SNP analyses 1 and 2 is 32, and annealing temperature for SNP analysis 3 is 59 °C. SNP reactions are carried out using the MegaBACE™ SNuPe Genotyping Kit using the following SNP primers: GATCACTGACGCAATCTTC, CAAGTTA GCCAACTTACAGA, and GGTAAGCCTGCACATCGTT, respectively.

3.2.3 Construction of a Genetic Linkage Map

JoinMap 3.0 [48] is used for map construction with maximum likelihood procedure. Among the markers, there are some anchor markers the positions of which are known, and this affects the choice of LOD (logarithm of odds) score. Primarily a LOD score of 9 is used but in some cases 8 (group 25), 10 (groups 5, 6, 7a, 7b, 11, 12), or 11 (groups 7c and 18) [22]. Map distances in centiMorgans (cM) are calculated by the Kosambi mapping function [49]. Maps for each linkage group are constructed using first- or second-round maps. The jump threshold is 5.0. A chi-square test is performed to determine the segregation of markers against an expected 1:1 ratio.

4 Notes

1. In our numerous experiments, anther culture response level of the control treatment varied between different donor plant sowings, although genotype was the same. This well known batch to batch variation can be assumed to be mainly dependent on seasonal variation in the growing conditions (early autumn and early to late spring being the best sowing times). However, other factors may also play a role, such as differences in the watering or fertilization of the plants, or possible disease infections during the growing period.
2. We used 1% 5-Superex solution (composition: 11% Nitrogen, 4% Phosphorus, 25% Potassium; or composition: 19% Nitrogen, 4% Phosphorus, and 20% Potassium, produced by company Kekkilä Finland), which was fed automatically to the watering system of the greenhouse tables. Extra dose once for young plantlets (three leaf stage) may be beneficial.
3. To give a high enough 2,4-D concentration (3–5 mg/L) was important in our experiments in order to get embryo-like structures which were capable to regenerate in to plants, although embryo-induction was got also with lower 2,4-D levels, and even without any growth regulators by applying only heat pretreatment. Small amount of Kinetin seemed to improve the quality of the induced ELS, promoting further regeneration to plants, although higher concentrations of cytokinins often caused browning of the anthers. Maltose gave better results in most of the tested genotypes, but sucrose might be better in some genotypes. We adopted W_{14} basic medium [24], but some genotypes may prefer MS-medium instead [25]. Ethylene-releasing component Ethephon™ and L-cysteine were added to improve induction of the regenerable-type ELSs, since despite good induction of ELS in many experiments, the regeneration was low or totally absent. Selection of cytokinin may be critical, since use of BAP in the induction

medium of cultivar Kolbu greatly increased the number of albinos compared to the same concentration of kinetin. We have also found preliminary evidence that the polyamine spermidine might increase induction of the regenerable-type ELSs. There are also other factors, which seem to have positive effects on either induction or regeneration phase, but which have not been confirmed due to the low induction levels, or just lack of time or labor.

4. Induction medium containing both a solid phase and a liquid phase was used to improve availability of the nutrients and hormones.
5. With one syringe filter about 50 ml of medium can be sterilized.
6. Try to use as fresh media as possible. Store for maximum 1 week.
7. It would have been beneficial to make DHs from the crossing parent first, before crossing them. But, that would have prolonged time to create mapping progeny too much in our project.
8. From the regenerated plants, five were rejected because they did not produce seeds, and five were potential heterozygotes according to the DNA-marker results. In addition, one regenerant was excluded for technical reasons so that DNAs fitted better on different formats (on plates and gels).
9. Use the main tillers if you want to optimize recovery of high number and quality of anthers.
10. Cold pretreatment can be 2 days shorter or longer, even 2 weeks in some genotypes may be beneficial for embryo induction. This flexibility helps to plan timetables. Also sowing anther-donor plants in intervals (e.g., 1 week) helps to scedule laborious anther isolation phase.
11. Sterilization of the panicles (which are cut out from the stem) with the 2% sodium hypochlorite (made from 10% NaOCl solution) with two to three drops of Tween 20 detergent, and rinsing three times with distilled water, may be necessary for tillers taken from the field conditions.
12. Developmental stage of flowers inside the spikelet varies according to where they are located in the panicle structure. Anther length and color can be used to reflect the developmental stage of microspores inside the anther. In general, the longer and greener the anther, the more the chance of microspores in the right developmental stage for anther culture. In addition, there are three sets of flowers in one spikelet of oat, and they are of different sizes. We excise anthers primarily from the most developed and biggest flowers.

13. In the induction phase, incubation is usually done in the dark. But in some genotypes dim light seems to be more beneficial (approximately 40 μmol m^2/s under a 16-h photoperiod in our experiments).
14. Also +25 °C is used successfully as induction temperature. Genotypes may react differently.
15. For some genotypes, low light in the induction phase seemed to be better than total darkness. This has been noticed earlier for recalcitrant genotypes of wheat also.
16. Ploidy level can be determined also by making chromosome preparations from the root tips, and use light microscope for counting the chromosome number. However, flow cytometry analysis is quick, reliable, and convenient.
17. It is important to check ploidy levels and treat haploid plants with colchicine, since also haploid oat plants grow and develop panicles in the greenhouse, but they do not produce seeds. There seems to be a genotypic variation in the rate of spontaneous chromosome doubling. In our experiments, common doubling rates were 15–30% of the regenerated plants, the majority being haploids. Also regeneration rate of the albino plants seemed to be genotype dependent, although some treatments increase their number.
18. Shorter and quicker DNA extraction methods can also be used for oat. However, we have encountered some problems after storing DNAs for some time, e.g., the marker bands get fainter and their size smaller evidently due to disruption of DNA. Therefore, we recommend a longer extraction method with a phenol–chloroform extraction which seems to be crucial.
19. Do not dry DNA for too long because it might prevent dissolving.
20. If there are problems in dissolving the precipitated DNA, it can be warmed to 60 °C for a couple of hours or even overnight.
21. DNAs of the progeny can be stored diluted on a 96-well plate, and then pipetted for PCR reactions on another 96-well plate using a multichannel pipette. However, DNAs start to degrade if they are too dilute. Therefore, it is better to have more concentrated DNA (e.g., 20 ng/μl) and to pipette less (1 μl) for PCR reactions than to use more diluted DNA (e.g., 5 ng/μl) and to pipette more (4 μl).
22. After testing the parents, PCR program for some primers can be modified slightly (e.g., increasing cycles) to get more distinct results when analyzing the progeny.
23. Reduced cycles compared to Li et al. [30] are used because otherwise peaks are too high in the capillary sequencer.

24. When microsatellites from other species are used, there are often problems in amplification and therefore PCR conditions have to be optimized, e.g., by lowering annealing temperature and increasing the number of cycles. However, this can lead to extra peaks, which make interpretation difficult. On the other hand, even when the microsatellites are amplified properly, the amount of polymorphism is usually low.
25. We noticed that marker bands amplify better when Failsafe D buffer is used instead of the buffer supplied by the polymerase manufacturer.

Acknowledgments

This work was supported by the Ministry of Agriculture and Forestry of Finland.

References

1. Forster BP, Thomas WTB (2005) Doubled haploids in genetics and plant breeding. Plant Breed Rev 25:57–88
2. Rines HW (1983) Oat anther culture: genotype effect on callus initiation and the production of a haploid plant. Crop Sci 23:268–272
3. Kiviharju E, Puolimatka M, Pehu E (1997) Regeneration of anther-derived plants of *Avena sterilis* L. Plant Cell Tiss Org Cult 48:147–152
4. Kiviharju E, Puolimatka M, Saastamoinen M, Hovinen S, Pehu E (1998) The effect of genotype on anther culture response of cultivated and wild oats. Agric Food Sci Finl 7:409–422
5. Kiviharju E, Pehu E (1998) The effect of cold and heat pretreatments on anther culture response of *Avena sativa* and *A. sterilis*. Plant Cell Tiss Org Cult 54:97–104
6. Kiviharju E, Tauriainen A (1999) 2,4-Dichlorophenoxyacetic acid and kinetin in anther culture of cultivated and wild oats and their interspecific crosses: plant regeneration from *Avena sativa* L. Plant Cell Rep 18:582–588
7. Kiviharju E, Laurila J, Lehtonen M, Tanhuanpää P, Manninen O (2004) Anther culture properties of oat × wild red oat progenies and a search for RAPD markers associated with anther culture ability. Agric Food Sci 13:151–162
8. Tuvesson S, Dayteg C, Hagberg P, Manninen O, Tanhuanpää P, Tenhola-Roininen T, Kiviharju E, Weyen J, Förster J, Schondelmaier J, Lafferty J, Marn M, Fledck A (2007) Molecular markers and doubled haploids in European plant breeding programmes. Euphytica 158(3):305–312
9. Slusarkiewicz-Jarzina A, Ponitka A (2007) The effect of physical medium state on anther culture response in polish cultivated oat (*Avena sativa* L.). Acta Biol Cracov Ser Bot 49:27–31
10. Ponitka A, Slusarkiewicz-Jarzina A (2009) A regeneration of oat androgenic plants in relation to induction media and culture conditions of embryo-like structures. Acta Soc Bot Pol 78(3):209–213
11. Rines HW, Riera-Lizarazu O, Nunez VM, Davis DW, Phillips RL (1997) Oat haploids from anther culture and from wide hybridizations. In: Jain SM, Sopory SK, Veilleux RE (eds) *In vitro* haploid production in higher plants, vol 4. Kluwer, Dordrecht, pp 205–221
12. Sun CS, Lu T-G, Söndahl MR (1991) Anther culture of naked oat and the establishment of its haploid suspension cell. Acta Bot Sin 33:417–420
13. Sidhu PK, Davies PA (2009) Regeneration of fertile green plants from oat isolated microspore culture. Plant Cell Rep 28:571–577
14. Ferrie AMR, Irmen KI, Beattie AD, Rossnagel BG (2014) Isolated microspore culture of oat (*Avena sativa* L.) for the production of doubled haploids: effect of pre-culture and post-culture conditions. Plant Cell Tiss Org Cult 116(1):89–96

15. Rines HW (2003) Oat haploids from wide hybridization. In: Maluszynski M, Kasha KJ, Forster BP, Szarejko I (eds) Doubled haploid production in crop plants. Kluwer, Dordrecht, pp 155–159
16. Sidhu PK, Howes NK, Aung T, Zwer PK, Davies PA (2006) Factors affecting oat haploid production following oat × maize hybridization. Plant Breed 125:243–247
17. Riera-Lizarazu O, Rines H, Phillips RL (1996) Cytological and molecular characterization of oat × maize partial hybrids. Theor Appl Genet 93(1-2):123–135
18. Kiviharju E, Puolimatka M, Saastamoinen M, Pehu E (2000) Extension of anther culture to several genotypes of cultivated oats. Plant Cell Rep 19:674–679
19. Kiviharju E, Moisander S, Laurila J (2005) Improved green plant regeneration rates from oat anther culture and the agronomic performance of some DH lines. Plant Cell Tiss Org Cult 81:1–9
20. Kiviharju EM (2009) Anther culture derived doubled haploids in oat. In: Forster BP, Jain SM, Touraev A (eds) Advances in Haploid Production in Higher Plants. Springer, Berlin, pp 171–178
21. Tanhuanpää P, Kalendar R, Schulman A, Kiviharju E (2008) The first doubled haploid linkage map for cultivated oat. Genome 51:560–569
22. Tanhuanpää P, Manninen O, Beattie A, Eckstein P, Scoles G, Rossnagel B, Kiviharju E (2012) An updated doubled haploid oat linkage map and QTL mapping of agronomic and grain quality traits from Canadian field trials. Genome 55:289–301
23. Tanhuanpää P, Manninen O, Kiviharju E (2010) QTLs for important breeding characteristics in the doubled haploid oat progeny. Genome 53:482–493
24. Ouyang JW, Jia SE, Zhang C, Chen X, Feng G (1989) A new synthetic medium (W_{14}) for wheat anther culture. Annual report 1987–1988. Institute of Genetics Academia Sinica, Beijing, pp 91–92
25. Murashige T, Skoog F (1962) A revised medium for rapid growth and bio assays with tobacco tissue culture. Physiol Plant 15:437–497
26. Dunwell JM (1985) Anther and ovary culture. In: Bright SWJ, Jones MKG (eds) Cereal tissue and cell culture. Nijhoff/Junk, Dordrecht, The Netherlands, pp 1–44
27. Immonen S, Tauriainen A, Manninen O (1999) Assesment of green regenerants from rye and triticale anther cultures. In: Clement C, Pacini E, Audran JC (eds) Anther and pollen: from biology to biotechnology. Springer, Berlin, pp 237–245
28. Bennett HD, Smith JB, Heslop-Harrison JS (1982) Nuclear DNA amount in angiosperms. Philos Trans R Soc Lond B 216:179–199
29. Poulsen GB, Kahl G, Weising K (1993) Abundance and polymorphism of simple repetitive DNA sequences in *Brassica napus* L. Theor Appl Genet 85:994–1000
30. Li CD, Rossnagel BG, Scoles GJ (2000) The development of oat microsatellite markers and their use in identifying relationships among *Avena* species and oat cultivars. Theor Appl Genet 101:1259–1268
31. Holland JB, Helland SJ, Shaporova N et al (2001) Polymorphism of PCR-based markers targeting exons, introns, promoter regions, and SSRs in maize and introns and repeat sequences in oat. Genome 44:1065–1076
32. Pal N, Sandhu JS, Domier LL et al (2002) Development and characterization of microsatellite and RFLP-derived PCR markers in oat. Crop Sci 42:912–918
33. Jannink J-L, Gardner SW (2005) Expanding the pool of PCR-based markers for oat. Crop Sci 45:2383–2387
34. Becher R (2007) EST-derived microsatellites as a rich source of molecular markers for oats. Plant Breed 126:274–278
35. Becker J, Heun M (1995) Barley microsatellites: allele variation and mapping. Plant Mol Biol 27:835–845
36. Liu ZW, Biyashev RM, Saghai Maroof MA (1996) Development of simple sequence repeat DNA markers and their integration into a barley linkage map. Theor Appl Genet 93:869–876
37. Ramsay L, Macaulay M, Degli Ivanissevich S et al (2000) A simple sequence repeat-based linkage map of barley. Genetics 156:1997–2005
38. Hackauf B, Wehling P (2002) Identification of microsatellite polymorphisms in an expressed portion of the rye genome. Plant Breed 121:17–25
39. Zietkiewicz E, Rafalski A, Labuda D (1994) Genome fingerprinting by simple sequence repeat (SSR)-anchored polymerase chain reaction amplification. Genomics 20:176–183
40. Kalendar R, Grob T, Regina M et al (1999) IRAP and REMAP: two new retrotransposon-based DNA fingerprinting techniques. Theor Appl Genet 98:704–711
41. Schulman AH, Gupta PK, Varshney RK (2004) Organization of retrotransposons and microsatellites in cereal genomes. In: Gupta PK,

Varshney RK (eds) Cereal genomics. Kluwer Academic Publishers, Dordrecht, The Netherlands, pp 83–118

42. Kalendar R, Schulman AH (2006) IRAP and REMAP for retrotransposon-based genotyping and fingerprinting. Nat Protoc 1:2478–2484
43. Vos P, Hogers R, Bleeker M et al (1995) AFLP: a new technique for DNA fingerprinting. Nucleic Acids Res 23:4407–4414
44. Li G, Quiros CF (2001) Sequence-related amplified polymorphism (SRAP), a new marker system based on a simple PCR reaction: its application to mapping and gene tagging in *Brassica*. Theor Appl Genet 103:455–461
45. Budak H, Shearman RC, Parmaksiz I et al (2004) Molecular characterization of Buffalograss germplasm using sequence-related amplified polymorphism markers. Theor Appl Genet 108:328–334
46. Tinker NA, Kilian A, Wight CP et al (2009) New DArT markers for oat provide enhanced map coverage and global germplasm characterization. BMC Genomics 10:39
47. Tanhuanpää P, Kalendar R, Laurila J, Manninen O, Schulman AH, Kiviharju E (2006) Generation of SNP markers for short straw in oat (*Avena sativa* L.). Genome 49:282–287
48. Van Ooijen JW, Voorrips RE (2001) Joinmap® 3.0, Software for the calculation of genetic linkage maps. Plant Research International, Wageningen, The Netherlands
49. Kosambi DD (1944) The estimation of map distance from recombination values. Ann Eugen 12:172–175

Chapter 7

Agrobacterium-Mediated Transformation of Leaf Base Segments

Sebastian Gasparis

Abstract

Agrobacterium-mediated transformation has become a routine method of genetic engineering of cereals, gradually replacing the biolistic protocols. Simple integration patterns of transgenic loci, decent transformation efficiency, and technical simplicity are the main advantages offered by this method. Here we present a detailed protocol for the production of transgenic oat plants by *Agrobacterium*-mediated transformation of leaf base segments. The use of leaf explants as target tissues for transformation and in vitro regeneration of transgenic plants may be a good alternative for genotypes which are not susceptible to regeneration from immature or mature embryos. We also describe the biochemical and molecular analysis procedures of the transgenic plants including a GUS histochemical assay, and Southern blot, both of which are optimized for application in oat.

Key words *Agrobacterium tumefaciens*, Transformation, Leaf base segment, Regeneration, GUS, Southern blot, DIG

1 Introduction

Historically oat has been one of the most important cereals used mainly as feed for livestock. In recent decades the importance of oats as a source of animal feed has dropped dramatically and has gradually been replaced by corn and barley. In regard to the excellent food value, oat has recently gained much popularity among consumers looking for health food products. The health-benefit properties of oat resulting from high amounts of desirable nutrients such as proteins, fiber, beta glucan, unsaturated fatty acids, minerals, and antioxidants have been well documented, making oat a valuable commodity in the food, pharmaceutical, and cosmetic industries. This simultaneously resulted in demand for increased grain quality.

Among many factors limiting the agricultural production and quality of oat products include environmental stresses such as drought and salinity stress. The breeding efforts are directed at

Sebastian Gasparis (ed.), *Oat: Methods and Protocols*, Methods in Molecular Biology, vol. 1536,
DOI 10.1007/978-1-4939-6682-0_7, © Springer Science+Business Media LLC 2017

improving stress tolerance, disease resistance, and other agronomic traits such as yield, lodging, and milling properties. These efforts can be effective only by combining recent biotechnology achievements with traditional breeding practices. Genetic engineering is the most important crop improvement strategy used to enhance agricultural productivity, and genetic transformation is the main tool for application of these strategies. Along with the recent advent of genome editing techniques, such as TALENS and CRISPR/Cas which have revolutionized genetic engineering, the development of efficient transformation protocols is particularly desired.

The number of reports of successful genetic transformation of oat is significantly lower as compared to other important cereal crops. Oat was first genetically modified in 1992 [1] by biolistic transformation of immature embryos. At this time it was the only method of transformation suitable for monocot species. Later it was applied to different types of cultivars and explants including immature embryos [2–6], mature embryos [7, 8], leaf base segments [9], shoot apical meristems [10–12], and immature embryo derived callus [13]. However, the application of biolistic transformation for genetic improvement is very limited because of the commonly known side effects of this method. The most important drawbacks are very complex transgene loci consisting of multiple copies [14], which are often rearranged and interspersed by genomic DNA [3, 15, 16] and lead to incorrect segregation patterns [17, 18] and low fertility [16]. In contrast, single or few transgene copy integration, stable transgene expression, good fertility, and more efficient transformation makes the *Agrobacterium*-mediated method more preferable for genetic modification of cereals [19–21]. The first successful *Agro*-transformation of oat was reported in 2008 [22], more than one decade later than rice [23], maize [24], wheat [25], or barley [26]. The protocol was developed in our laboratory for three cultivars, two husked and one naked, and for two types of explants, immature embryos, and leaf base segments. The transformation procedures were designed to test three combinations of *A. tumefaciens* strains and vectors, LBA4404(pTOK233), EHA101(pGAH), and AGL1(pGreen) and three different selection systems: the *hpt* gene conferring resistance to hygromycin (pTOK233), the *nptII* gene conferring resistance to kanamycin (pGAH and pGreen), and the *bar* gene conferring resistance to phosphinothricin (pGreen). The highest transformation efficiency—12.3% for immature embryos and 8.3% for leaf base segments was obtained using the pTOK233 vector and kanamycin selection. The transformation of immature embryos with the pGreen vector containing the kanamycin selection cassette resulted in a lower, but still satisfactory efficiency of 5.3%. It is worth stating that two commonly used vector families pGreen (now available as pBract) and pCambia are suitable for *Agrobacterium*-mediated transformation of oat using our procedure. The pCAMBIA1201 vector containing the hygromycin resistant gene (*hpt*) has been successfully applied to transform different cultivars of oat [27].

Immature and mature embryos are the most widely used target tissues for the genetic transformation of cereals mainly due to their excellent in vitro regeneration ability. However, application of this type of explant is time and space consuming, and dependent on long term cultivation of donor plants in controlled conditions. Another problem is related to the occasionally occurring prolonged tissue culture required for the formation of the embryogenic callus, which increases the risk of somaclonal variations. For these reasons, the development of an efficient, embryo-independent regeneration system is particularly desired. For oat, another two types of explants—shoot apical meristems [10–12] and leaf base segments [9, 28–31] were adopted for in vitro regeneration and genetic transformation of different cultivars. Both types of tissues provided a high frequency of regeneration and reproducibility, and the regenerated plants showed high fertility and genomic stability. Another advantage of using leaf explants is that they can be easily collected from germinated seeds which are available the entire year as opposed to donor plants for immature embryos which should be grown under strictly controlled conditions. It greatly reduces the time of the transformation cycle and saves space in growth chambers.

Here we present a method for the production of transgenic oat plants by *Agrobacterium*-mediated transformation of leaf base segments. The procedure was developed in our laboratory several years ago [22]. Transgenic plants were obtained from cultivar Bajka after selection on kanamycin and the transformation frequency was 8.3 %, estimated as the percentage of the selected, independent callus lines which gave rise to at least one transgenic plant.

Further, we describe the molecular and biochemical methods to confirm transgene integration in the transformed plants. They include procedures for nonradioactive DIG Southern blot analysis and a GUS histochemical assay.

2 Materials

2.1 Plant Material

Mature seeds of the Polish oat cultivar Bajka (*see* **Note 1**).

2.2 Agrobacterium tumefaciens Strain and Vector

A. tumefaciens strain LBA4404 carrying the pTOK233 binary vector [23] is used. The pTOK233 vector contains a selection cassette with the *nptII* gene under the CaMV35S promoter. The *nptII* gene confers resistance to the aminoglycoside antibiotic kanamycin. Additionally, pTOK233 carries the *gus* reporter gene under the CaMV35S promoter (*see* **Note 2**).

2.3 Stock Solutions

2.3.1 Stock Basal Media Components

1. MS macrosalts [32] (10×): 16.5 g/L NH_4NO_3, 4.4 g/L $CaCl_2 \cdot 2H_2O$, 3.7 g/L $MgSO_4 \cdot 7H_2O$, 1.7 g/L KH_2PO_4, 19 g/L KNO_3 (*see* **Note 3**).
2. MS microsalts [32] (1000×): 6.3 mg/L H_3BO_3, 0.025 mg/L $CoCl_2 \cdot 6H_2O$, 0.025 $CuSO_4 \cdot 5H_2O$, 16.9 mg/L $MnSO_4 \cdot 4H_2O$, 0.83 mg/L KI, 0.25 mg/L $Na_2MoO_4 \cdot 2H_2O$, 8.6 mg/L $ZnSO_4 \cdot 7H_2O$.
3. Iron chelate (200×): 7.45 g/L Na_2EDTA, 5.55 g/L $FeSO_4 \cdot 7H_2O$.
4. Gamborg [33] B5 vitamin stock (100×): 10 g/L myoinositol, 0.1 g/L nicotinic acid, 0.1 g/L pyridoxine HCl, 1.0 g/L thiamine HCl. Store in 20 mL aliquots at −20 °C. If precipitation occurs, warm the solution in a water bath at 37 °C before use.
5. 2,4-dichlorophenoxyacetic acid (2,4-D) (1000×): Dissolve 0.15 g in 10 mL of 70% ethanol; when dissolved bring up to a final volume of 100 mL. Store at 4 °C.
6. 6-benzylaminopurine (BAP) (1000×): Dissolve 0.1 g in 10 mL ddH_2O, add a few drops of 1 N NaOH for complete dissolution, and bring up to a final volume of 100 mL with ddH_2O. Filter-sterilize and store 1 mL aliquots at −20 °C.
7. 3-indolylacetic acid (IAA) (1000×): Dissolve 0.2 g in a final volume of 100 mL of 70% ethanol. Store 1 mL aliquots at −20 °C.
8. Biotin (1000×): 1 mg/mL stock in ddH_2O. Filter-sterilize and store 1 mL aliquots at −20 °C.
9. Acetosyringone: 20 mM stock in 70% ethanol. Dissolve 0.0785 g in a final volume of 20 mL ethanol. Store 0.5 mL aliquots at −20 °C.

2.3.2 Antibiotics

1. Hygromycin B (1000×): 50 mg/mL stock in ddH_2O. Filter-sterilize and store 1 mL aliquots at 4 °C.
2. Kanamycin sulfate (1000×): 50 mg/mL stock in ddH_2O. Filter-sterilize and store 1 mL aliquots at −20 °C.
3. Timentin (ticarcillin disodium/potassium clavulanate) (1000×): 150 mg/mL stock in ddH_2O. Filter-sterilize and store 1 mL aliquots at 4 °C.
4. Rifampicin (1000×): 50 mg/mL stock in DMSO. Store 0.5 mL aliquots at −20 °C in darkness.

2.4 Culture Media

2.4.1 Media for *Agrobacterium tumefaciens*

1. AB medium [34] (*see* **Note 4**):
   (a) AB salts stock (20×): 20 g/L NH_4Cl, 6 g/L $MgSO_4 \cdot 7H_2O$, 3 g/L KCl, 0.2 g/L $CaCl_2 \cdot 2H_2O$, 0.05 g/L $FeSO_4 \cdot 7H_2O$, autoclave at 121 °C for 20 min.

(b) AB buffer (20×): 60 g/L KH_2PO_4, 20 g/L NaH_2PO_4, autoclave at 121 °C for 20 min.

(c) 5 g/L glucose: dissolve 0.5 g in 90 mL H_2O, filter-sterilize.

To prepare 100 mL of AB medium add 5 mL of AB buffer and 5 mL of AB salts stock to 90 mL sterile glucose solution.

2. Inoculation medium: Liquid MSB1 medium (without Gelrite; see below) combined with acetosyringone to a final concentration of 200 μM. Acetosyringone is added to sterile MSB1 medium before its use for inoculation.

2.4.2 Media for Oat In Vitro Cultures

1. Germination medium: Half-strength (0.5×) MS macrosalts and microsalts, 2.5 mL/L iron chelate stock, 20 g/L sucrose, 3 g/L Gelrite. Adjust pH to 5.8 with 1 N KOH, autoclave at 121 °C for 20 min.
2. Cocultivation medium MSB1 (solid): 1× MS macrosalts and microsalts, 5 mL/L iron chelate stock, 10 mL/L Gamborg B5 vitamin stock, 3 mg/L 2,4-D, 30 g/L sucrose, 3 g/L Gelrite. Adjust pH to 5.8 with 1 N KOH, autoclave at 121 °C for 20 min.
3. Callus induction medium: Same as MSB1; after autoclaving add 150 mg/L Timentin (for *Agrobacterium* suppression and elimination) and 50 mg/L kanamycin (for selection).
4. Regeneration medium MIB: 1 × MS macrosalts and microsalts, 5 mL/L iron chelate stock, 10 mL/L Gamborg B5 vitamin stock, 30 g/L sucrose, 3 g/L Gelrite. Adjust pH to 5.8 with 1 N KOH, autoclave at 121 °C for 20 min. After autoclaving add 1 mg/L BAP, 0.2 mg/L IAA, and 150 mg/L Timentin.
5. Rooting medium: same as germination medium with appropriate antibiotics. After autoclaving add 50 mg/L kanamycin and 150 mg/L Timentin.

2.5 Reagents for the GUS Staining Assay

1. X-GlcA (5-bromo-4-chloro-3-indolyl-β-D-glucuronic acid, cyclohexylammonium salt): 40 mM stock in DMF. Dissolve 20 mg in 1 mL DMF (*N,N*-dimethylformamide). Store at −20 °C in darkness.
2. Sodium phosphate buffer: 50 mM $NaHPO_4$ pH 7.0, 0.1% Triton X-100.
3. Chlorophyll destaining solution: 96% ethanol mixed with 99% acetic acid in a ratio of 3:1.

2.6 Reagents for Southern Hybridization

1. Kits for DIG labeling and detection (*see* **Note 5**).
2. Anti-digoxigenin-AP Fab fragments.
3. Blocking reagent.

4. Hybridization buffer: DIG Easy-Hyb granules or standard hybridization buffer.
5. Denaturing buffer: 1.5 M NaCl, 0.5 M NaOH.
6. Neutralizing buffer: 0.5 M Tris–HCl pH 7.0, 1.5 M NaCl.
7. 20× SSC: 0.3 M sodium citrate, 3 M NaCl; adjust pH to 7.0; autoclave.
8. Maleic acid buffer: 0.1 M maleic acid, 0.15 M NaCl, adjust pH to 7.5 (for 1 L dissolve about 8 g NaOH pellets then carefully adjust to pH 7.5 with 1 and 0.1 N NaOH solutions), autoclave.
9. Washing buffer: maleic acid buffer with 0.3% (v/v) Tween 20. Prepare from sterile maleic acid buffer, do not autoclave.
10. Blocking solution: 10% (w/v) solution in maleic acid buffer. Add 10 g of blocking reagent to 100 mL of maleic acid buffer. Stir while heating (do not allow to boil) until fully dissolved (no granules visible); do not autoclave.
11. Detection buffer: 0.1 M Tris–HCl pH 9.5, 0.1 M NaCl.
12. Low stringency washing buffer: 2× SSC, 0.1% SDS.
13. High stringency washing buffer: 0.5 × SSC, 0.1% SDS.

2.7 Other Reagents and Supplies

1. 0.1% $HgCl_2$ solution for sterilization of kernels (*see* **Note 6**).
2. Tween 20 surfactant.
3. Parafilm M sealing tape, 100 mm width.
4. Petri dishes 80 mm.
5. Soil—we use universal flower soil mixed with sand in a ratio of 10:1.
6. Plastic pots 140–160 mm in diameter with drainage holes.
7. Yellow and blue greenhouse sticky traps 10 × 20 cm.
8. Fertilizer—for the most convenient application use a liquid, concentrated, multipurpose fertilizer which can be diluted and applied directly by watering.
9. Nylon membranes, positively charged.
10. Whatman 3MM filter paper.
11. Hybridization bottles or bags.
12. Film for chemiluminescent detection.

3 Methods

3.1 Transformation Procedures

3.1.1 Obtaining Leaf Base Explants

1. Use mature, dry seeds of the chosen oat cultivar. Remove hulls by hand.
2. Place approximately 200 kernels in a 250 mL Erlenmeyer flask for surface sterilization. From this point all further steps should be done in a laminar flow hood in sterile conditions.

Fig. 1 Transformation of oat: (**a**) 4-day-old seedlings on MS medium, (**b**) 4–5 mm leaf base segments on cocultivation medium, (**c**) 2-week-old explants on callus induction medium, (**d**) embryogenic callus on regeneration medium, (**e**) regenerating transformed plantlets, (**f**) rooting and selection of transformed plants on kanamycin-containing rooting medium

3. Cover the flask with sterile gauze. Pour 70% ethanol to cover the kernels and wash them for 1 min.
4. Discard the ethanol, wash the kernels in sterile ddH_2O and discard, add one drop of Tween-20, and pour in 0.1% $HgCl_2$ solution. Wash the kernels for 3–4 min.
5. Discard the $HgCl_2$ solution (*see* **Note 6**) and wash the kernels three times with sterile ddH_2O. After the third wash, leave the kernels in water for 15 min.
6. Place the kernels on germination medium; apply six kernels per plate. Seal the plates with Parafilm.
7. Germinate the seeds at 22 °C in darkness for 4–5 days to obtain seedlings with 2–3 cm long coleoptiles (Fig. 1a).
8. Cut out the 3–5 mm leaf base segment just above the embryo. Place 6–7 segments per plate on MSB1 cocultivation medium, seal the plates with Parafilm (Fig. 1b). Transfer the plates to the culture room for a 2-day preculture under the following conditions: temperature 22 °C day/night, and a 16-h photoperiod under 65–70 μmol/m^2/s light.

3.1.2 Inoculation and Cocultivation of Explants with *Agrobacterium tumefaciens*

1. Prepare the *Agrobacterium* culture suspension 2 days before the expected inoculation of leaf explants. Inoculate 20 mL of AB medium with 100 μL of *Agrobacterium* glycerol stock (stored at −80 °C) in a 100 mL Erlenmeyer flask and add 50 mg/L hygromycin and 50 mg/L rifampicin. Shake the culture on a horizontal shaker at 28 °C, 140 rpm.

2. Grow the culture until an OD_{600} value of 0.5–1 is reached (usually 36–48 h). Centrifuge the *Agrobacterium* suspension in Falcon tubes for 8 min at room temperature, 4500 × *g*.
3. Discard the supernatant and resuspend the pellet in 10 mL of MSB inoculation medium with the addition of 200 μM acetosyringone. Gently shake the tube and incubate the culture at room temperature until the OD_{600} reaches 0.8–2. Use this suspension to inoculate the previously precultured leaf segments.
4. Put approximately 30 leaf segments into a sterile 50 mL Erlenmeyer flask and pour in 10 mL of *Agrobacterium* suspension. Gently shake the flask on a horizontal shaker for about 20 min at room temperature.
5. Pipette out the *Agrobacterium* suspension. Place the leaf segments on fresh MSB1 cocultivation medium, six segments per plate. Incubate the explants in the culture room at 22 °C in darkness for 2 days (*see* **Note 7**).

3.1.3 Induction of Embryogenic Calli, Selection, and Regeneration

1. After cocultivation, transfer the inoculated leaf segments to a fresh callus induction medium containing 50 mg/L kanamycin and 150 mg/L Timentin. Place six segments per plate.
2. Incubate the plates in the culture room under the same conditions as described in Subheading 3.1.1, **step 8**. For the first 2 weeks the plates should be covered with 1–2 layers of filter paper to maintain twilight (Fig. 1c).
3. Every fourth to fifth day cut the elongating leaves or coleoptiles to keep the appropriate size of 4–5 mm of the leaf base segment (*see* **Note 8**).
4. After 4–5 weeks of culture, dark yellow somatic embryo clusters should emerge on the calli which survived the selection (Fig. 1d). A plantlet or a group of them developed from each callus is considered as an independent transgenic event.
5. Transfer 2–3 embryonic calli onto a separate plate containing fresh MIB regeneration medium with 50 mg/L kanamycin and 150 mg/L Timentin. Remove the remaining leaf segment.
6. The somatic embryos will develop into plantlets during the next 2–4 weeks of culture. The leaves which emerge in the early stage of regeneration are visible as green dots (Fig. 1e). Remove each developing embryo/plantlet from the callus and place it on the surface of the MIB medium. Use a stereoscopic microscope if necessary.

3.1.4 Rooting and Acclimatization of the Transgenic Plants

1. For the rooting and growth of transgenic plantlets use glass jars, approx. 12–16 cm high, or similar in vitro containers. Pour the rooting medium into the jars; about 0.25 of the jar volume is enough for root development.

2. Transfer 2–3 cm plantlets from the MIB medium to the rooting medium with a selection agent. Carefully remove the surrounding callus. Seal the jars with Parafilm. Rooting is maintained in the culture room under the same conditions (*see* **Note 9**).
3. Plants with well-developed roots can be put into the soil (Fig. 1f). Use pots at least 140 mm in diameter with drainage holes (*see* **Note 10**).
4. Plants from in vitro culture must be adapted to the new conditions, especially the lower humidity. For acclimatization, cover each plant in the pot with a transparent jar. Alternatively, you can use commercially available humidity domes. Keep the plants covered for 1–2 weeks (*see* **Note 11**).
5. Grow the plants in the growth chamber until maturity under the following conditions: temperature 18 °C day/12 °C night, and a 16-h photoperiod with 350 μmol/m^2/s light intensity (*see* **Note 12**).

3.1.5 Collection and Storage of Transgenic Seeds

1. Mature transgenic plants should be transferred to the greenhouse. Higher temperature and lower humidity are better for seed maturation.
2. Cut the panicles; collect only mature, dry, and healthy grains. Remove the lemmas and paleas. Only bare seeds should be stored.
3. For optimal storage conditions, the grain moisture should be below 14 % and the temperature below 20 °C. Long-term storage requires temperatures below 4 °C.

3.2 GUS Histochemical Assay

The GUS assay is a method which allows the detection of the expression of the *UidA* reporter gene in transformed tissues. *UidA* encodes the enzyme β-glucuronidase (GUS) which hydrolyzes glucuronide substrates to colored or fluorescent products that can be visualized by histochemical or fluorogenic assays.

In the GUS histochemical assay, the colorless substrate 5-bromo-4-chloro-3-indolyl-β-D-glucuronic acid is processed to a dark blue/indigo product 3-hydroxy-5-bromo-4-chloroindole (Fig. 2).

1. Prior to each use, prepare fresh sodium phosphate buffer and add X-GlucA stock solution to a final concentration of 2 mM. Pipette 150–200 μL of this buffer into Eppendorf tubes or microtiter plates.
2. Take a small tissue fragment (*see* **Note 13**) and submerge it completely in the buffer. For better infiltration of tissues by the buffer, apply a vacuum for 15 min.
3. Incubate the samples at 37 °C for 24 h in darkness.

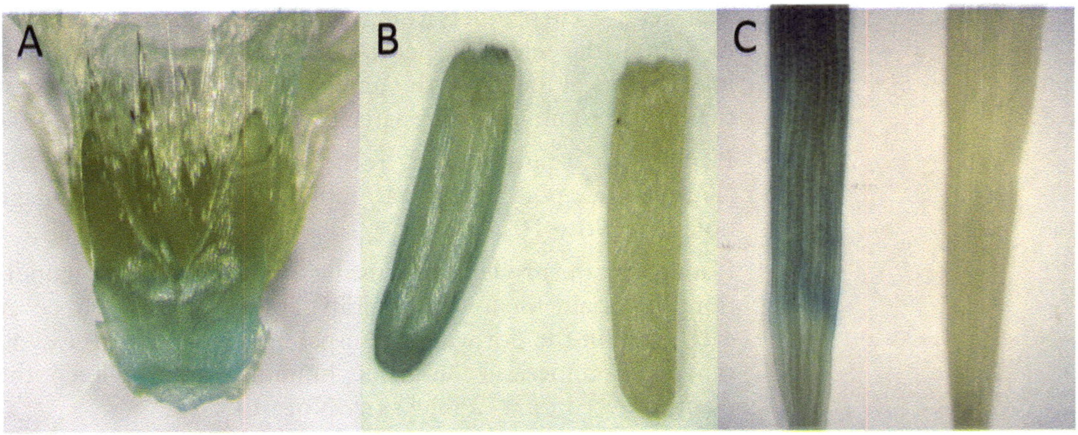

Fig. 2 GUS expression in various organs of transgenic oat plants. (**a**) Immature floret, (**b**) coleoptile section (*left*), control sample (*right*), (**c**) leaf section (*left*), control sample (*right*)

4. Remove the sodium phosphate buffer and wash the samples in destaining solution for 30 min at room temperature. If necessary, repeat the washing several times until all of the chlorophyll has been removed.
5. Samples can be stored for longer time in 75 % ethanol in the dark at room temperature.

3.3 Southern Blot Analysis of Transgenic Plants

The procedure is optimized for nonradioactive Southern hybridization with digoxigenin (DIG). For DNA labeling, hybridization, and detection we used reagents provided by Roche. The manufacturer's protocols were optimized by us for the specific application to oat and other cereals with large genomes.

3.3.1 Preparation of Genomic DNA and Capillary Transfer

1. Use freshly isolated, pure genomic DNA from transgenic T_0 or T_1 plants and non-transgenic control plants. Check the purity and integration of the isolated DNA in a spectrophotometer and on an agarose gel respectively.
2. Digest from 20 to 40 μg of genomic DNA with the HindIII restriction enzyme for 12–16 h (*see* **Note 14**).
3. Digest the plasmid DNA of your transformation vector. Use the enzyme which excises the fragment containing the transgene (we digested the pTOK233 vector with the KpnI enzyme to isolate the fragment containing the sequence of the *nptII* gene). Use at least 60 pg of digested plasmid DNA for blotting.
4. Fractionate the digested genomic DNA and plasmid DNA on a 0.7% agarose gel in TBE buffer. Use a DIG-labeled DNA marker for each gel. Apply low voltage to resolve the gel for at least 12 h.

5. Denature the DNA by gently shaking the gel in denaturation buffer for 30 min, briefly wash the gel in ddH_2O and neutralize for 30 min in neutralization buffer, next wash the gel in 10× SSC for another 30 min.
6. Perform a standard capillary transfer with at least 300 mL 10× SSC (*see* **Note 15**). Transfer the DNA overnight to a positively charged nylon membrane.
7. Briefly wash the membrane in 2× SSC to remove all gel residues. Place the membrane between two sheets of 3MM paper and bake it at 120 °C for 30 min (or according to the manufacturer's instructions).

3.3.2 DNA Labeling and Hybridization

1. Label the fragment of your transgene with DIG using the PCR DIG Probe Synthesis Kit (Roche) according to the manufacturer's instructions. Take 5–10 μL from the PCR reaction for electrophoresis on an agarose gel to estimate the amount of amplified probe (Fig. 3).
2. Pre-hybridize the membrane at 42 °C for 30 min with the proper volume of DIG Easy Hyb buffer (10 mL/100 cm^2 of membrane).
3. In the meantime denature the DIG-labeled probe (*see* **Note 16**) by boiling it for 5 min in a water bath and immediately cooling on ice. Add the denatured probe to the pre-warmed (42 °C) DIG Easy Hyb buffer (3.5 mL/100 cm^2 of membrane).

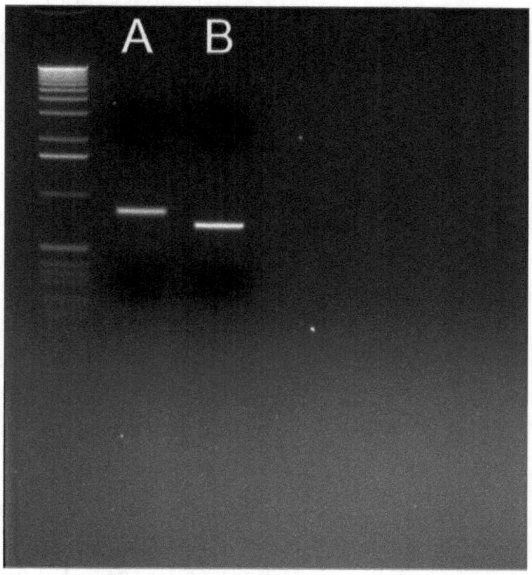

Fig. 3 DIG labeling of the 700 bp fragment of *nptII* in the PCR reaction. The labeled fragment (**a**) migrates slower on the gel than the control sample (**b**). 5 μL of PCR product (**a**) was taken to prepare the probe mixture

4. After the pre-hybridization discard the DIG Easy Hyb buffer and add the probe mixture. Hybridize the membrane at 42 °C overnight (12–16 h).

3.3.3 Chemiluminescent Detection

1. After hybridization, wash the membrane twice with low stringency buffer for 5 min at room temperature, followed by 2 × 15 min in high stringency buffer at 65 °C in a hybridization oven.
2. Place the membrane on a plastic tray (DNA side upward!) and wash it for 1–3 min in washing buffer.
3. Incubate the membrane in 100 mL blocking solution for 30 min at room temperature.
4. Prepare the antibody solution: centrifuge the vial with the Anti-DIG AP Fab fragments for 5 min at 10,000 × *g*, pipette 2 μL of antibodies from the upper part of the solution and add it to 20 mL of blocking solution. Incubate the membrane in the antibody solution with gentle shaking for 30 min at room temperature.
5. Wash the membrane 2 × 15 min in washing buffer and then incubate for 2–5 min in 20 mL of detection buffer.
6. Place the membrane on a plastic wrap (Saragold or equivalent) and add a few drops of CSPD (included in the detection kit). Immediately cover the membrane with another sheet of plastic wrap to evenly spread the CSPD on the whole membrane surface.
7. Remove the excess CSPD and air bubbles by rolling the membrane with a glass pipette. Incubate the wrapped membrane at 37 °C for 10 min in the dark.
8. Place the membrane in a photographic cassette and expose to chemiluminescent film at room temperature for 1 h (Fig. 4) (*see* **Note 17**).

4 Notes

1. Many oat cultivars have been successfully regenerated from leaf base explants and it seems that regeneration susceptibility is comparable with mature or immature embryos. However, we recommend testing the susceptibility of the cultivar to be genetically transformed.
2. We have obtained higher transformation rates with the pTOK233 vector both in the case of immature embryos and leaf base segments. Taking into account that this vector was developed in the 1990s (it is large and difficult for cloning), we suggest using the pBract or pCAMBIA vector systems, which

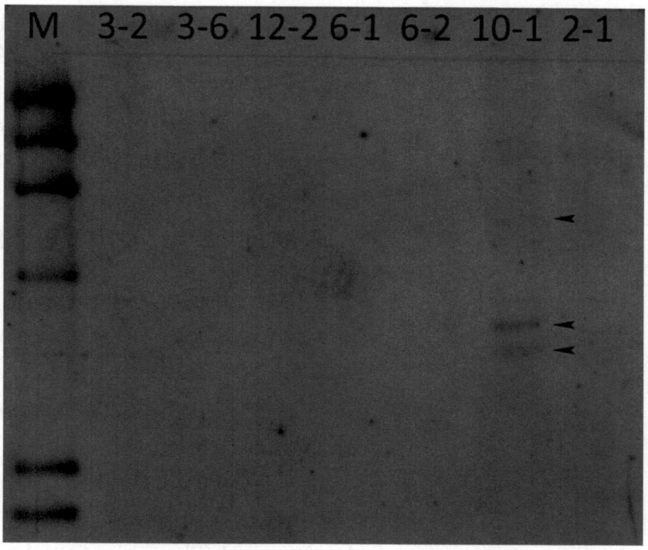

Fig. 4 Southern blot analysis of the T_1 progeny of transformed oat plants. 40 μg of genomic DNA was digested with *HindIII* and hybridized with a 700-bp *nptII* PCR amplified and DIG labeled probe. *Arrowheads* indicate the bands of the detected transgenes. The chemiluminescent film was exposed for 14 h

are small, easy to clone, and intended for cereal transformation. If you apply a different strain and vector remember to use the appropriate bacterial medium and antibiotics for your system.

3. Longer storage of the MS salts stock solution is not recommended as it may be easily infected or some components may precipitate. Therefore, we recommend using premixed MS salts as a ready to use powder which is dissolved during preparation of the media.
4. AB minimal medium is recommended for growth of *Agrobacterium* strain LBA4404. For AGL1 strain use MG/L medium [35]. It is not recommended to grow *Agrobacterium* strains on rich media and at temperature above 28 °C as it may result in a loss of binary vector by bacteria cells.
5. The procedure for the Southern hybridization was optimized by us for use with the DIG labeling and detection system provided by Roche. The following kits and reagents are used in the procedure: PCR DIG Probe Synthesis Kit (11636090910), DIG Luminescent Detection Kit (11363514910), DIG Easy Hyb Granules (11796895001), Anti-Digoxigenin-AP, Fab fragments (11093274910), Nylon Membranes, positively charged (11417240001), Lumi-Film Chemiluminescent Detection Film (11666657001). In Europe these products are now distributed by Sigma-Aldrich.

6. $HgCl_2$ is highly toxic. Read the instructions in the Material Safety Data Sheet before use. Always use personal protective equipment while handling this chemical. Waste should be stored in glass bottles and recycled in accordance with local regulations. Do not release the waste to the environment.
7. Overgrowth of *Agrobacterium* may prevent callus formation and regeneration. Control the growth of *Agrobacterium* during the subsequent 1–2 days after inoculation. The explants should be surrounded by faint *Agrobacterium* slime. Discard the explants which are overgrown (fully covered) by bacteria. Use a suspension with a lower OD_{600} for inoculation.
8. Take notice that the callus is formed only at the lower end of the leaf segment (from the embryo side). Therefore, cut the elongating leaves at the opposite (upper) side of the leaf segment.
9. Note that the rooting medium also contains kanamycin for plant selection and Timentin for *Agrobacterium* suppression. It has often been reported that some non-transgenic plants may "escape" selection. Based on our experience, this additional selection step greatly eliminates the non-transgenic escapes.
10. Plants with at least 3–4 roots might be put into the soil. It is very important to remove all medium residues from the roots to prevent fungal infections.
11. Usually 1 week of acclimatization is enough for well-developed plants. Longer growth periods in humid conditions may cause fungal infections. Plants with weak, thin leaves require more time or they may dry out completely. Keep them covered until their appearance is similar to a typical plant.
12. Growth chambers should be controlled systematically for pests and fungal diseases. We use yellow and blue sticky monitoring cards to control pests. Thrips and fungal gnats are the most common pests in our growth chambers. If necessary we use Talstar™ and Nomolt™ to prevent them. To minimize the risk of diseases, always keep the growth chambers clean. Remove dead leaves and avoid excessive watering.
13. We performed GUS assays for various types of tissues and organs and the higher GUS activities were observed in florets and young leaves (*see* Fig. 2). In turn, no signal was detected in roots and seeds. It has been reported that endogenous GUS activity may occur in some tissues. For each assay include a non-transgenic plant as a control sample to eliminate false-positives with endogenous GUS activity.
14. Based on our experience, 20 μg of genomic DNA is the minimum amount in which the transgene is detectable. On the other hand, applying more than 40 μg of DNA to the blot is not recommended.

15. We suggest performing downward capillary transfer with 10× SSC which gave us better results than the vacuum blot. Transfer in alkali buffers (e.g., 0.4 M NaOH) is not suitable for DIG-labeled molecular weight markers.
16. Take the labeled probe directly from the PCR reaction. The exact amount of the probe applied for hybridization should be estimated by analyzing the PCR product on an agarose gel. Usually, 2 μL/mL of hybridization buffer is enough if the band of the amplified product is clearly visible on the gel (*see* Fig. 3). If the band is very faint, use up to 4 μL of probe per mL of hybridization buffer. Applying higher amounts may result in a strong background on the membrane or unspecific hybridization of the probe.
17. A strong hybridization signal should be visible after 40–60 min of exposure to the film. Based on our experience, the film should be exposed for several hours to detect the signal. We suggest checking the film after 1 h. If no signal is visible and the film is transparent (without a dark background), one can expose another film for 12–16 h (*see* Fig. 4).

Acknowledgment

This research was supported by the Statutory Funds of Plant Breeding and Acclimatization Institute—National Research Institute, No. 1-1-01-4-04.

References

1. Somers DA, Rines HW, Gu W, Kaeppler HF, Bushnell WR (1992) Fertile, transgenic oat plants. Nat Biotechnol 10(12):1589–1594. doi:10.1038/nbt1292-1589
2. Torbert KA, Rines HW, Somers DA (1995) Use of paromomycin as a selective agent for oat transformation. Plant Cell Rep 14(10):635–640
3. Pawlowski WP, Somers DA (1998) Transgenic DNA integrated into the oat genome is frequently interspersed by host DNA. Proc Natl Acad Sci U S A 95(21):12106–12110. doi:10.1073/pnas.95.21.12106
4. Torbert KA, Gopalraj M, Medberry SL, Olszewski NE, Somers DA (1998) Expression of the Commelina yellow mottle virus promoter in transgenic oat. Plant Cell Rep 17(4):284–287
5. Kuai B, Perret S, Wan SM, Dalton SJ, Bettany AJE, Morris P (2001) Transformation of oat and inheritance of bar gene expression. Plant Cell Tissue Organ Cult 66(2):79–88. doi:10.1023/A:1010613704824
6. Perret SJ, Valentine J, Leggett JM, Morris P (2003) Integration, expression and inheritance of transgenes in hexaploid oat (Avena sativa L.). J Plant Physiol 160(8):931–943. doi:10.1078/0176-1617-00880
7. Torbert KA, Rines HW, Somers DA (1998) Transformation of oat using mature embryo-derived tissue cultures. Crop Sci 38(1):226–231
8. Cho MJ, Jiang W, Lemaux PG (1999) High-frequency transformation of oat via microprojectile bombardment of seed-derived highly regenerative cultures. Plant Sci 148(1):9–17. doi:10.1016/S0168-9452(99)00082-5
9. Gless C, Lörz H, Jähne-Gärtner A (1998) Transgenic oat plants obtained at high efficiency by microprojectile bombardment of leaf base segments. J Plant Physiol 152(2–3):151–157. doi:10.1016/s0176-1617(98)80126-0

10. Zhang S, Cho MJ, Koprek T, Yun R, Bregitzer P, Lemaux PG (1999) Genetic transformation of commercial cultivars of oat (Avena sativa L.) and barley (Hordeum vulgare L.) using in vitro shoot meristematic cultures derived from germinated seedlings. Plant Cell Rep 18(12):959–966
11. Maqbool B, Zhong H, El-Maghraby Y, Ahmad A, Chai B, Wang W, Sabzikar R, Sticklen B (2002) Competence of oat (Avena sativa L.) shoot apical meristems for integrative transformation, inherited expression, and osmotic tolerance of transgenic lines containing hva1. Theor Appl Genet 105(2–3):201–208. doi:10.1007/s00122-002-0984-3
12. Cho MJ, Choi HW, Okamoto D, Zhang S, Lemaux PG (2003) Expression of green fluorescent protein and its, inheritance in transgenic oat plants generated from shoot meristematic cultures. Plant Cell Rep 21(5):467–474. doi:10.1007/s00299-003-0582-0
13. Kaeppler HF, Menon GK, Skadsen RW, Nuutila AM, Carlson AR (2000) Transgenic oat plants via visual selection of cells expressing green fluorescent protein. Plant Cell Rep 19(7):661–666
14. Pawlowski WP, Somers DA (1996) Transgene inheritance in plants genetically engineered by microprojectile bombardment. Mol Biotechnol 6(1):17–30. doi:10.1007/Bf02762320
15. Svitashev SK, Somers DA (2001) Genomic interspersions determine the size and complexity of transgene loci in transgenic plants produced by microprojectile bombardment. Genome44(4):691–697.doi:10.1139/gen-44-4-691
16. Choi HW, Lemaux PG, Cho MJ (2001) High frequency of cytogenetic aberration in transgenic oat (Avena sativa L.) plants. Plant Sci 160(4):763–772. doi:10.1016/s0168-9452(01)00391-0
17. Kohli A, Leech M, Vain P, Laurie DA, Christou P (1998) Transgene organization in rice engineered through direct DNA transfer supports a two-phase integration mechanism mediated by the establishment of integration hot spots. Proc Natl Acad Sci U S A 95(12):7203–7208
18. Leggett JM, Perret SJ, Harper J, Morris P (2000) Chromosomal localization of cotransformed transgenes in the hexaploid cultivated oat Avena sativa L. using fluorescence in situ hybridization. Heredity 84(1):46–53. doi:10.1046/j.1365-2540.2000.00627.x
19. Dai SH, Zheng P, Marmey P, Zhang SP, Tian WZ, Chen SY, Beachy RN, Fauquet C (2001) Comparative analysis of transgenic rice plants obtained by Agrobacterium-mediated transformation and particle bombardment. Mol Breed 7(1):25–33. doi:10.1023/A:1009687511633
20. Travella S, Ross SM, Harden J, Everett C, Snape JW, Harwood WA (2005) A comparison of transgenic barley lines produced by particle bombardment and Agrobacterium-mediated techniques. Plant Cell Rep 23(12):780–789. doi:10.1007/s00299-004-0892-x
21. Zalewski W, Orczyk W, Gasparis S, Nadolska-Orczyk A (2012) HvCKX2 gene silencing by biolistic or Agrobacterium-mediated transformation in barley leads to different phenotypes. BMC Plant Biol 12:206. doi:10.1186/1471-2229-12-206
22. Gasparis S, Bregier C, Orczyk W, Nadolska-Orczyk A (2008) Agrobacterium-mediated transformation of oat (Avena sativa L.) cultivars via immature embryo and leaf explants. Plant Cell Rep 27(11):1721–1729. doi:10.1007/s00299-008-0593-y
23. Hiei Y, Ohta S, Komari T, Kumashiro T (1994) Efficient transformation of rice (Oryza sativa L.) mediated by Agrobacterium and sequence analysis of the boundaries of the T-DNA. Plant J 6(2):271–282
24. Ishida Y, Saito H, Ohta S, Hiei Y, Komari T, Kumashiro T (1996) High efficiency transformation of maize (Zea mays L.) mediated by Agrobacterium tumefaciens. Nat Biotechnol 14(6):745–750. doi:10.1038/nbt0696-745
25. Cheng M, Fry JE, Pang SZ, Zhou HP, Hironaka CM, Duncan DR, Conner TW, Wan YC (1997) Genetic transformation of wheat mediated by Agrobacterium tumefaciens. Plant Physiol 115(3):971–980
26. Tingay S, McElroy D, Kalla R, Fieg S, Wang M, Thornton S, Brettell R (1997) Agrobacterium tumefaciens-mediated barley transformation. Plant J 11(6):1369–1376. doi:10.1046/j.1365-313X.1997.11061369.x
27. Oraby H, Ahmad R (2012) Physiological and biochemical changes of CBF3 transgenic oat in response to salinity stress. Plant Sci 185:331–339. doi:10.1016/j.plantsci.2012.01.003
28. Chen Z, Zhuge Q, Sundqvist C (1995) Oat leaf base: tissue with an efficient regeneration capacity. Plant Cell Rep 14(6):354–358. doi:10.1007/BF00238596
29. Chen H, Xu G, Loschke DC, Tomaska L, Rolfe BG (1995) Efficient callus formation and plant regeneration from leaves of oats (Avena sativa L.). Plant Cell Rep 14(6):393–397. doi:10.1007/BF00238604
30. Gless C, Lörz H, Jähne-Gärtner A (1998) Establishment of a highly efficient regeneration system from leaf base segments of oat (Avena sativa L.). Plant Cell Rep 17(6):441–445. doi:10.1007/s002990050422
31. Nuutila AM, Villiger C, Oksman-Caldentey K-M (2002) Embryogenesis and regeneration of green plantlets from oat (Avena sativa L.) leaf-base

segments: influence of nitrogen balance, sugar and auxin. Plant Cell Rep 20(12):1156–1161. doi:10.1007/s00299-002-0467-7

32. Murashige T, Skoog F (1962) A revised medium for rapid growth and bio assays with tobacco tissue cultures. Physiol Plant 15(3):473–497. doi:10.1111/j.1399-3054.1962.tb08052.x
33. Gamborg OL, Miller RA, Ojima K (1968) Nutrient requirements of suspension cultures of soybean root cells. Exp Cell Res 50(1):151–158
34. Chilton MD, Currier TC, Farrand SK, Bendich AJ, Gordon MP, Nester EW (1974) Agrobacterium tumefaciens DNA and PS8 bacteriophage DNA not detected in crown gall tumors. Proc Natl Acad Sci U S A 71(9):3672–3676
35. Garfinkel DJ, Nester EW (1980) Agrobacterium-tumefaciens mutants affected in crown gall tumorigenesis and octopine catabolism. J Bacteriol 144(2):732–743

Part III

Quality Assessment

Chapter 8

Chromatographic Methods to Evaluate Nutritional Quality in Oat

Gracia Montilla-Bascón, Corey D. Broeckling, Owen A. Hoekenga, Elena Prats, Mark Sorrells, and Julio Isidro-Sánchez

Abstract

Oats (*A. sativa* L.) have an important and positive role in human diet and health. The health benefits of oats are attributed to its multifunctional characteristic and nutritional profile, being an important source of soluble dietary fiber, well-balanced proteins, unsaturated fatty acids, vitamins, essential minerals, and a good source of natural antioxidants. These antioxidants include the avenanthramides (Avns) and avenalumic acids, which are unique to oats among cereals. High-performance liquid chromatography allows a simultaneous quantification of free amino acids and biogenic amines in oat samples as their OPA/FMOC-CL (*o*-phthalaldehyde/9-fluorenylmethoxycarbonyl chloride) derivatives. In addition, an ultra-performance liquid chromatography/mass spectrometry method was developed to quantify and characterize avenanthramides contained in oat samples.

Key words Oats, Amino acids, Biogenic amines, Avenanthramides, HPLC, UPLC-MS

1 Introduction

Oats (*A. sativa* L.) are recognized for their beneficial nutritional properties for human health. Compared to other cereals, this grain has high content of soluble fiber, betaglucans, proteins, and lipids, as well as a unique source of avenanthramides (Avns) (phenolics alkaloids) [1, 2]. For breeding programs, characterization of the nutritional profile is essential to enhance oats for nutritional and medical purposes.

The development of modern analytical chemistry methods such as high and ultra performance liquid chromatography has made it possible to separate, identify and quantify different compounds in a mixture at ng/g or lower levels [3]. In addition, coupling a mass spectrophotometer (MS) to chromatographic separation such as gas chromatography (GC) and liquid chromatography (LC) has improved sensitivity and accuracy of determination and quantification of the different molecules and compounds contained in a sample relative to chromatography alone [4].

Sebastian Gasparis (ed.), *Oat: Methods and Protocols*, Methods in Molecular Biology, vol. 1536,
DOI 10.1007/978-1-4939-6682-0_8,

High-performance liquid chromatography (HPLC) is a powerful technique to separate and detect compounds and their derivatives important for nutritional quality, including primary and secondary amino acids and biogenic amines by ultraviolet (UV) fluorescent detection using derivatizing reagents OPA and FMOC-CL simultaneously [5, 6]. OPA is one of the most widely used derivatizing reagents to detect fluorescent derivatives of free amino acids and biogenic amines. However, OPA cannot react with secondary amino acids such as proline and hydroxyproline, which are derivatized using FMOC-CL reagent [5, 6]. Amino acid derivatization is required prior to HPLC analysis to replace active hydrogens on OH, NH_2, and SH polar functional groups with a nonpolar moiety. Thereby, amino acids become less reactive and their chromatographic behavior is improved.

HPLC offers the selectivity, sensitivity, and resolution necessary to separate and identify the fluorescent derivatives of both amino acids and biogenic amines [7, 8]. Ultra-performance liquid chromatography coupled with mass spectrometry (UPLC-MS) is suitable for large-scale analysis of metabolism in biological systems [9]. This analytical approach allows separation and detection of the highly rich biochemistry of plants, including secondary metabolite groups such as Avns [10]. UPLC-MS is a powerful method to analyze targeted metabolites important to biological functions. Alternatively, non-targeted metabolomics analysis by UPLC-MS is a broad and unbiased approach to enable detection of as many metabolites as possible. However, a targeted metabolomics analysis has the advantage of having a predetermined panel of known metabolites and thus it facilitates the identification of new metabolites [11]. The use of this method can save of time, effort and money since a guided analysis for each metabolite is not necessary each time. Known metabolites can be quantified using standards and new metabolites can be characterized at the same time.

In this chapter, an improved methodology to determine free primary, secondary amino acids, and biogenic amines in oats by HPLC method using combined derivatization with OPA/FMOC-CL reagents has been developed to enhance speed and accuracy of quantification for oats samples. Furthermore, a UPLC-MS method has been developed to separate, identify and quantify the three major forms of avenanthramides: Avn A, Avn B, and Avn C, as well as other putative avenanthramides contained in oat samples.

2 Materials

2.1 Plant Materials

A diverse collection of spring oat accessions was used in this experiment to analyze primary and secondary amino acids and biogenic amines [6]. Seeds were planted at two field trial locations near Ithaca, NY and harvested when ripe. Seeds were dried at 21 °C in a dryer chamber to remove excess moisture.

2.2 Amino Acids Quantification

2.2.1 Equipment

1. 1.5 and 2 mL safe-lock Eppendorf tubes or equivalents.
2. Laboratory scale (100 g×0.1 mg).
3. Vortex mixer.
4. 42 °C heat-block with shaker (ThermoMixer F1.5; Eppendorf).
5. HPLC with a fluorescent detector and a 20 μL loop injector (Agilent 2100 Series, Santa Clara, CA, USA).
6. Column oven at 42 °C.
7. Column Merck Lichro-CART®250-4 Superspher®100 RP-18 endcapped (25 cm×4.6 mm; 5 mm particles).
8. Guard column LiChrospher® 100 RP-18 (5 μm) LiChroCART® 4-4.
9. 25 μL HPLC Syringe with stainless steel needle.
10. 0.5 mm filters.

2.2.2 Reagents and Solutions

1. Free amino acid standards stock solutions (22): prepare 1.5 mg/L solutions of aspartic acid (Asp), glutamic acid (Glu), asparagine (Asn), serine (Ser), glutamine (Gln), threonine (Thr), glycine (Gly), histidine (His), alanine (Ala), γ-aminobutyric acid (GABA), tyrosine (Tyr), valine (Val), methionine (Met), tryptophan (Trp), β-phenylalanine (Phe), isoleucine (Ile), leucine (Leu), lysine (Lys), citrulline (Cit), ornithine (Orn), 4.53 mg/mL solution of arginine (Arg) and 25.44 mg/mL solution of proline (Pro). Prepare stock solutions 100× for each amino acid, except for proline that needs a 1000× stock, to be used in the calibration curve (*see* **Note 1**).
2. Biogenic amine standards (10): prepare 4.2 mg/mL solution of ethanolamine (ETA), methylamine (METILA), ethylamine (ETILA), histamine (HISTA), tyramine (TIR), β-phenylethylamine (FEN), tryptamine (TRY), isoamilamine (ISO), cadaverine (CAD), and putrescine (PUT). Prepare stock solutions 100× for each biogenic amine to be used in the calibration curve.
3. Internal standard, 5.86 mg/L norvaline (Nval).
4. 0.4 M borate buffer: dissolve 6.18 g of boric acid in 250 mL of distilled water, and then use 0.5 M NaOH to get the desired pH in each case, 10.4 or 10.6.
5. OPA reagent: dissolve 51.2 mg of *o*-phthalaldehyde in 1.5 mL of ethanol, add 80 μL of 3-mercaptopropionic acid and 1750 mg of potassium thiocyanate (KSCN) and complete with 0.4 M borate buffer (pH 10.4) until a final volume of 50 mL (*see* **Note 2**). Divide the reagent into aliquots and store at −20 °C; under these conditions it is stable for 2 months.
6. FMOC-CL reagent: dissolve 25 mg FMOC-CL in 10 mL of acetonitrile obtaining a final concentration of 2.5 g/L. Divide into aliquots and store at −15 °C being stable during several months (*see* **Note 3**).

7. Mobile phase A: 99.67% 20 mM sodium acetate buffer, 0.018% (v/v) triethylamine, 0.3% (v/v) tetrahydrofurane, and 0.010% of a EDTA solution 0.4% (m/v). For 1 L of solution: 996.72 mL 20 mM sodium acetate buffer, 180 μL triethylamine, 3 mL tetrahydrofuran, and 100 μL EDTA 0.4% (m/v) and adjust the pH to 7.2 with 0.2% acetic acid.
8. Mobile phase B: 20% (v/v) of 100 mM sodium acetate buffer pH 6.0, 40% (v/v) of acetonitrile, 40% (v/v) of methanol, and 0.018% of triethylamine. For 1 L of solution: 200 mL 100 mM sodium acetate buffer pH 6.0, 400 mL acetonitrile, 400 mL methanol, and 180 μL triethylamine.

2.3 Avenanthramide Relative Quantification

2.3.1 Equipment

1. 5 mL Next Advance tubes or equivalents.
2. Laboratory scale.
3. Refrigerated vortex.
4. Next Advance Bullet Blender bead beater or equivalent.
5. Waters Acquity UPLC with Waters Xevo TOF MS.
6. Column oven at 50 °C.
7. Waters Acquity UPLC HSS T3 C18 column (1.8 μM, 1.0 × 100 mm).
8. 1.0 L graduated cylinder.

2.3.2 Reagents and Solutions

1. Avenanthramide standards: Avenanthramide A, B, and C are available from ReseaChem (Burgdorf, Switzerland). Prepare stock solutions of the three avenanthramide standards 1.0 mg/mL in 80% methanol.
2. Internal standard (optional).
3. Buffer A. Rinse graduated cylinder thoroughly with LC-MS grade water. Fill cylinder to 500 mL with LC-MS grade water and add 1.0 mL LC-MS grade Formic Acid. Gently swirl to mix. Add LC-MS grade water to 1.0 L and transfer to clean LC solvent reservoir.
4. Buffer B. Rinse graduated cylinder thoroughly with LC-MS grade acetonitrile. Fill cylinder to 500 mL with LC-MS grade acetonitrile and add 1.0 mL LC-MS grade Formic Acid. Gently swirl to mix. Add LC-MS grade acetonitrile to 1.0 L and transfer to clean LC solvent reservoir.
5. Extraction solution methyl-t-butyl ether (MTBE)–methanol–water: Dispense 600 mL of methyl-t-butyl ether (MTBE) in a bottle, add 300 mL of methanol and finally add 100 mL of LC-MS grade water. Gently swirl to mix.

3 Methods

3.1 Setting Up

Chromatography methods and TOF MS data are both very reproducible when used correctly. Chromatographic methods are very sensitive to issues such as small particles, air bubbles, pH changes, external temperature changes, etc. Ensure that your solvents are fresh, the lines are well purged, and that all leak tests pass before starting the analyses. On the MS, ensure that the instrument is tuned appropriately and calibrated.

1. Clean HPLC and UPLC-MS by running methanol through pumps and tubes during 2 h to remove possible particles and microorganisms, which could obstruct the flow. During this process, the pumps should be opened to purge the machine.
2. Before using, filter the mobile phase A to amino acids quantification with a 0.5 mm filter to avoid blockage of the HPLC circuit due to possible KSCN undissolved residues.
3. Equilibrate each method for at least 1 h by flushing the system with the initial elution conditions; 95 % mobile phase A and 5 % mobile phase B with a constant flow of 0.7 mL/min to separate amino acids by HPLC and 100% Buffer A with 0.2 mL/min constant flow rate to separate avenanthramides by UPLC-MS.

3.2 Standard Curves and Properties

3.2.1 Amino Acids and Amines

1. Calculate the calibration standard curves for each compound injecting 10 μL of different concentration solutions: 100×, 50×, 25×, and 1×, obtained from the stock solutions. The standard concentration average ranged from 1.1 to 11 mg/L for all the amino acids except for arginine, which ranged from 4.53 to 45.29 mg/L and for proline, ranging from 25.44 to 254.4 mg/L. Standard concentrations for amines ranged from 0.15 to 4.20 mg/L. Prepare the calibration solutions immediately before injection, and include the internal standard in all of them in a concentration of 5.86 mg/L (*see* **Note 4**). Follow the same method used for the analysis of experimental samples.
2. Analyze the retention time and spectrum for each compound to identify perfectly each peak of interest. First, inject each standard solution separately in order to characterize chromatographically each one individually. Further, inject all standards together in a complete injection to study possible changes in their chromatographic characteristics when they are analyzed at the same time (*see* **Note 5**). Several complete injections were carried out containing the specific concentrations used to develop the calibration curves (100×, 50×, 25× and 1×). Standards retention times obtained in our analyses have been included as a reference (Table 1).

Table 1
Retention times of amino acids and biogenic amines detected in oat samples

Compound	Retention time
Aspartate	3.2
Glutamate	3.6
Asparagine	6.2
Serine	6.8
Glutamine	7.6
Threonine	8.3
Glycine	8.8
Histidine	9.2
Citrulline	10.6
Alanine	12.9
Arginine	14.1
γ-aminobutyric acid	14.4
Tyrosine	16.6
Ornithine	18.5
Valine	24.5
Methionine	26
Norvaline	27.4
Tryptophan	32.3
β-phenylethylamine	33.9
Isoleucine	36
Histamine	37.7
Leucine	46.27
Lysine	65.2
Tyramine	77
Putrescine	78
Proline	79.7

3.2.2 Avenanthramides

A dilution series should be prepared covering the full dynamic range of the instrument (typically 3–4 orders of magnitude) (*see* **Note 6**). The molecular ions observed in positive ionization mode for the three standards are protonated ($[M+H]+ = 300.086$,

330.097, 316.082, respectively, for avenanthramides A, B, and C) and sodiated ($[M+Na]+=322.069$, 352.080, 338.065, respectively, for avenanthramides A, B, and C). The base peak ions for these signals represents the ionized phenolic in-source fragment (147.044, 177.055, and 163.039, respectively, for avenanthramides A, B, and C). These ions are also useful diagnostics for recognizing other less abundant avenanthramides. The retention order on the Waters T3 column under acidic acetonitrile conditions is Avenanthramide C (4.71 min), Avenanthramide A (5.12 min), and Avenanthramide B (5.32 min). If adapting the assay to a tandem quadrupole instrument, $[M+H]+$ to base peak ion should provide optimal SRM sensitivity.

3.3 Amino Acids Determination

Dehull seeds using a mechanical dehuller (Codema LLC, Maple Grove, MN, USA), and select only intact grains for analyses. Weigh out approximately 15 g of dehulled oats in a 15 mL polycarbonate tube and grind with three 7/16″ stainless steel ball bearings (OPS Diagnostics, Lebanon, NJ, USA) at room temperature during 2 min at 600 rpm to get a fine powder (*see* **Note 7**). Then pass samples through a 0.5 mm filter to remove any hull remnants. Analyze three biological and three technical repetitions of each genotype.

3.3.1 Extraction

Homogenize 30 mg of oat flour in 300 μL 0.1 M HCl to get the seed extract. Incubate the extraction for 5 min at RT and vortex each minute for a short time. Then, centrifuge sample at 3000 rpm/min (1008 rcf/min) during 5 min at RT and collect supernatant in a fresh 1.5 mL Eppendorf tube and hold on ice until derivatization or immediately freeze at −20 °C to use later.

3.3.2 Derivatization of Amino Acids

Mix 40 μL of acidic oat extracts with 200 μL sodium borate buffer (0.4 M, pH 10.6) and 200 μL OPA reagent both at RT. Agitate for 20 s and add 40 μL FMOC-CL reagent to the mixture, then homogenize again for 15 s. Place the final reaction in a heat block at 42 °C for 2 min with gentle agitation.

3.3.3 Analytical Conditions

Inject 20 μL of the final reaction solution into the HPLC loop manually. Amino acids are separated using a binary grade method from mobile phase A and mobile phase B to resolve amino acid derivatives with an extension of 140 min (Table 2). Initial and final conditions are the same for re-equilibrating the column after each injection.

The UV detector should be set at an excitation/emission wavelength of 340/450 nm for the first 115 min, to better detect major amino acids. The UV detector should be changed to shorter wavelengths (237/340 nm) for the final 25 min, to better detect proline and minor amino acids.

Table 2
Amino acids and biogenic amines elution gradient

Time (min)	% Eluent A	% Eluent B	Flow (mL/min)
0	95	5	0.7
2	90	10	0.7
20	77	23	0.7
20.5	77	23	0.9
40	77	23	0.9
45	75	25	0.9
45.5	75	25	0.8
52.5	75	25	0.8
53	75	25	0.9
53.5	75	25	0.7
70	50	50	0.7
70.5	50	50	0.4
85	43	57	0.4
85.5	43	57	0.6
105	40	60	0.6
110	20	80	0.7
110.5	20	80	0.5
117	20	80	0.5
120	0	100	1.2
135	0	100	1.2
136	95	5	0.7
140	95	5	0.7

3.4 Avenanthramides Determination

3.4.1 Extraction

Homogenize 300 mg of oats flour in 3 mL of cold methyl-t-butyl ether (MTBE)–methanol–water (6:3:1 by volume) to obtain the seed extracts and mix samples at moderate speed for 1 h at 4 °C with a vortex mixer (*see* **Note 8**). To encourage separation into aqueous and organic phases, add 750 μL of cold LC-MS grade water to enable phase separation, mix briefly, and centrifuge at 3200 × *g* at 4 °C for 10 min to separate phases. Avenanthramides are aqueously soluble, so the 1.0 mL sample is drawn from the lower aqueous phase to a clean autosampler vial for subsequent analysis.

Table 3
Avenanthramides elution gradient

Time (min)	% Buffer A	% Buffer B	Flow (mL/min)
0.0	100	0	0.2
1	100	0	0.2
13	2	98	0.2
16	2	98	0.2
16.05	100	0	0.2
20	100	0	0.2

3.4.2 Analytical Conditions

Inject 10 μL of the sample extract for each sample (*see* **Note 9**). If concentration adjustment is necessary—evaporate solvent under nitrogen, and resuspend in 50% methanol 50% water at 25% the original volume. Compounds are separated using a binary gradient from buffer A (water, 0.1% formic acid) to buffer B (acetonitrile, 0.1% formic acid) with an extension of 20 min and a 0.2 mL/min constant flow rate (Table 3). The column and samples are held at 50 °C and 5 °C, respectively. The column eluent was infused into a Waters Xevo G2 TOF-MS with an electrospray source in positive mode, scanning 50–1200 m/z at 0.2 s per scan. The capillary voltage is held at 2200 V, source temp at 150 °C, and nitrogen desolvation temp at 350 °C with a flow rate of 800 L/h.

4 Notes

1. All solvents should be of analytical grade and used without further purification. For LC-MS, use LC-MS grade solvents and prepare all solutions using ultrapure water to reduce experimental error with chromatographic separation. Prepare and store all reagents at recommended conditions (unless indicated to be a fresh preparation).
2. OPA: Be sure that KSCN is completely dissolved after agitation as sometimes this reagent is in suspension but not totally dissolved. Depending on the number of samples that are going to be analyzed per day, single-use aliquots of 1, 1.5, or 2 mL should be prepared to use each day to waste the least amount of reagent.
3. FMOC-CL should also be divided into single-use aliquots appropriate to the number of samples per day that will be analyzed. FMOC-CL is reactive, such that it is recommended that aliquots not be stored at −4 °C for more than 6 months.

4. The range for calibrating each compound can be modified depending on method and instrument conditions, but at least four different concentration measurements are necessary for each standard in order to obtain a good curve fit. Nevertheless, it is crucial to add the internal standard in each injection and in equal concentration for every sample and standard analyzed. The internal standard Norvaline has stable analytical behavior and allows operators to check if the retention time and concentration appear correctly across all chromatography experiments.
5. Chromatographic characteristics of each standard, sample or substance, can change depending on the chromatographic system used for its analysis and within the same system when any circumstance changes. For that reason, the chromatographic characterization of each compound is the most important step in determining and quantifying compounds. It is necessary to carefully characterize samples and standard behavior to ensure accurate quantification. An accurate characterization of each standard acting alone and jointly with others is absolutely necessary as secondary peaks appear due to the interactions with the OPA and FMOC-CL derivatization reagents. Most of the amino acids produce one derivative, but some of them, especially asparagine, glutamine, and tryptophan, produce multiple derivatives [12]. The secondary peaks belonging to these derivatives can hide or modify the intended peaks produced by amino acids of highest interest. If you cannot identify each amino acid in the complete injection a modification in the reaction conditions may be implemented. Several modifications can be made to the derivatization reaction, such as, lowering the temperature, changing the reaction time, changing the flow or the eluents concentration, depending which amino acids need to be separated. For example, if tryptophan is especially important for your analysis, increasing the reaction time will enhance the separation of the preferred tryptophan derivative. For that reason, an accurate characterization of the interesting derivatized amino acid forms is essential, including excitation/emission spectrum, retention time, chromatogram, etc. If necessary, a mass spectrometer can be used to identify each derivative with higher confidence.
6. The exact concentrations of the calibration solutions are going to be determined based on the sensitivity of your instrument. You should obtain the minimum concentration detected by your UPLC and use the minimum value in the concentration curves.
7. The grinding time is hardness-dependent and it is important to prevent samples from overheating during the grinding process since heat can significantly affect some oat components, such as lipids. The hardness of oat grains is genotype-dependent, so, each one needs to be tested in order to know the minimum

grinding time. We recommend grinding for 2 min at 600 rpm and allowing the sample to cool before grinding 1 min more, if needed.

8. Methyl-t-butyl ether (MTBE)–methanol–water reagent does not need to be prepared daily, but it is advisable to prepare it weekly and preserve at 4 °C until used. You can prepare aliquots in bottles containing the daily volume necessary to prevent heating of the reagent.
9. Set the autosampler to 4 °C before injections, it is important not to break the cold chain during the sample preparation period until its injection.

References

1. Butt MS, Tahir-Nadeem M, Khan MKI et al (2008) Oat: unique among the cereals. Eur J Nutr 47:68–79
2. Singh R, De S, Belkheir A (2013) Avena sativa (Oat), a potential neutraceutical and therapeutic agent: an overview. Crit Rev Food Sci Nutr 53:126–144
3. Carrasco-Pancorbo A, Nevedomskaya E, Arthen-Engeland T et al (2009) Gas chromatography/atmospheric pressure chemical ionization-time of flight mass spectrometry: analytical validation and applicability to metabolic profiling. Anal Chem 81:10071–10079
4. Pitt JJ (2009) Principles and applications of liquid chromatography-mass spectrometry in clinical biochemistry. Clin Biochem Rev 30:19–34
5. Herbert P, Barros P, Ratola N et al (2000) HPLC determination of amino acids in musts and port wine using OPA/FMOC derivatives. J Food Sci 65:1130–1133
6. Herbert P, Santos L, Alves A (2001) Simultaneous quantification of primary, secondary amino acids, and biogenic amines in musts and wines using OPA/3-MPA/FMOC-Cl fluorescent derivatives. J Food Sci 66:1319–1325
7. Kelly MT, Blaise A, Larroque M (2010) Rapid automated high performance liquid chromatography method for simultaneous determination of amino acids and biogenic amines in wine, fruit and honey. J Chromatogr A 1217:7385–7392
8. Lehtonen P (1996) Determination of amines and amino acids in wine—a review. Am J Enol Vitic 47:127–133
9. Heuberger AL, Broeckling CD, Kirkpatrick KR et al (2014) Application of nontargeted metabolite profiling to discover novel markers of quality traits in an advanced population of malting barley. Plant Biotechnol J 12:147–160
10. De Vos RCH, Moco S, Lommen A et al (2007) Untargeted large-scale plant metabolomics using liquid chromatography coupled to mass spectrometry. Nat Protoc 2:778–791
11. Broeckling CD, Heuberger AL, Prenni JE (2013) Large scale non-targeted metabolomic profiling of serum by ultra performance liquid chromatography-mass spectrometry (UPLC-MS). J Vis Exp 73:e50242
12. Molnar-Perl I, Katona ZF (2000) GC-MS of amino acids as their trimethylsilyl/t-butyldimethylsilyl derivatives: in model solutions III. Chromatographia 51:S228–S236

Chapter 9

Determination of T-2 and HT-2 Toxins in Oats and Oat-Based Breakfast Cereals by Liquid-Chromatography Tandem Mass Spectrometry

Veronica M.T. Lattanzio and Michelangelo Pascale

Abstract

This protocol specifies an accurate and sensitive method for the determination of T-2 and HT-2 toxins content in oats and oat-based foods using liquid chromatography with tandem mass spectrometry detection. T-2 and HT-2 toxins are extracted from the test material with a mixture of acetonitrile and water. The filtered extract is dried, reconstituted with a mixture of methanol and water, then purified on a polymeric solid-phase extraction cartridge. Toxins are finally eluted from the column with methanol and quantified by reversed phase high performance liquid chromatography with tandem mass spectrometry detection.

Key words *Fusarium* toxins, T-2 toxin, HT-2 toxin, Oats, Liquid chromatography, Mass spectrometry, Solid phase extraction

1 Introduction

T-2 toxin (T-2) and HT-2 toxin (HT-2) are two of the most toxic type A trichothecenes produced mainly by *Fusarium sporotrichioides*, *F. poae*, and *F. langsethiae* that may develop on a variety of cereal grains, especially in cold climate regions or during wet storage conditions [1–4]. A detailed report on toxicity data of T-2 and HT-2 has been recently produced on behalf of EFSA (the European Food Safety Authority) [5]. T-2 is a potent inhibitor of protein synthesis and, at higher concentrations, of DNA and RNA synthesis. The toxicity of HT-2 has been less investigated, nevertheless due to the fact that T-2 is rapidly metabolized to HT-2 in vivo, it is widely accepted that the toxicity of T-2 in vivo includes that of HT-2 [1, 5, 6]. Results from the European project "SCOOP" (task 3.2.10), finalized in 2003, showed a low incidence of contamination by T-2 (20% out of 3490 analyzed samples) and HT-2 (14% out of 3032 analyzed samples) in cereals, including oats, maize, wheat, barley, and rye [2]. More recent data have shown

Sebastian Gasparis (ed.), *Oat: Methods and Protocols*, Methods in Molecular Biology, vol. 1536,
DOI 10.1007/978-1-4939-6682-0_9, © Springer Science+Business Media LLC 2017

oats to be the cereal most susceptible to T-2 and HT-2 contamination with incidence and concentration of toxins depending on the crop year [3, 4, 7, 8]. Maximum indicative levels for the sum of T-2 and HT-2, in cereals and cereal products, have been recently set by the European Commission [9]. In particular, for unprocessed oats a maximum indicative level of 1000 μg/kg has been established for the sum of T-2 and HT-2, whereas for oats intended for direct human consumption, oat bran and flaked oats the maximum indicative level is 200 μg/kg [9].

Liquid chromatography coupled with tandem mass spectrometry (LC–MS/MS) is currently widely used for the simultaneous determination of trichothecene mycotoxins, including T-2 and HT-2, in cereals and derived products [10]. Besides high sensitivity, tandem MS detection also provides the highest degree of certainty in analyte identification [10, 11].

In the present chapter, an LC-MS/MS method for the determination of T-2 and HT-2 toxins in oats and oat-based products is described. The method was developed and validated within a project funded by the UK Food Standards Agency (Project Number C03056), aimed at obtaining improved analytical methods for T-2 and HT-2 toxins to be validated for use in a range of food matrices, including oats and relevant processed products.

2 Materials

During the analysis, unless otherwise stated, use only reagents of recognized analytical grade. Solvents for analysis shall be of HPLC or better quality, and only double-distilled water or water of at least grade 2 as defined in EN ISO 3696 shall be used for sample extraction. Suitable precaution and protection measures need to be taken when carrying out working steps with harmful chemicals (*see* **Note 1**).

2.1 Sample Extraction

1. Analytical balance, capable to measure 10 mg.
2. Filter paper (pore size 20–25 μm, or similar).
3. Acetonitrile.
4. Extraction solvent mixture: mix 84 parts per volume of acetonitrile with 16 parts per volume of water.

2.2 Extract Cleanup

1. Graduated pipette (5 mL).
2. Solvent evaporator: heating block with air or nitrogen gas supply.
3. 10 mL Pyrex® test tubes with caps.
4. Solid-phase extraction (SPE) columns: Oasis® HLB column (polymeric column filled with [poly(divinylbenzene-co-*N*-vinylpyrrolidone)], with a capacity of 3 mL, containing 60 mg sorbent.

5. Vacuum manifold, to accommodate solid-phase extraction (SPE) columns.
6. Vials of 4 mL with caps.
7. Calibrated microliter syringe(s) or microliter pipette(s) (100–1000 μL).
8. SPE conditioning solution: mix 3 parts per volume of methanol with 7 parts per volume of water.
9. SPE washing solution: mix 55 parts per volume of methanol with 45 parts per volume of water.

2.3 LC-MS/MS Analysis

1. LC-MS/MS apparatus comprising the following:
   - Solvent delivery system. Capable of delivering a binary gradient at flow rates appropriate for the analytical column in use with sufficient accuracy (*see* **Note 2**).
   - Auto liquid sampler. Capable of injecting an appropriate volume of injection solution with sufficient precision (*see* **Note 2**). Cross contamination should be below 0.1 %.
   - Analytical column. Any column can be used provided that they ensure a baseline resolution of the T-2 and HT-2 peaks from all other peaks (*see* **Note 3**).
   - Column oven. Capable of maintaining a constant temperature above any variability caused by fluctuations in the room temperature (*see* **Note 2**).
   - Mass spectrometer. An instrument capable of ionization of T-2 and HT-2 toxins in positive ion mode, performing Selected Reaction Monitoring (SRM) with a sufficiently wide dynamic range. Any ionization source giving sufficient yield may be employed.
2. LC syringe filters (Nylon, 0.45 or 0.22 μm, depending on LC system requirements).
3. Autosampler vials with caps.
4. LC grade methanol.
5. Ammonium acetate, for mass spectrometry, $c(CH_3COONH_4) \geq 99.0\,\%$.
6. LC mobile phase: solvent A (5 mM CH_3COONH_4 in water): dissolve 0.385 g CH_3COONH_4 in 1 L of water; solvent B (5 mM CH_3COONH_4 in methanol): dissolve 0.385 g CH_3COONH_4 in 1 L of methanol.
7. Injection solvent: mix 4 parts per volume of solvent A with 6 parts per volume of solvent B.
8. T-2 and HT-2 toxins, either in form of crystals or film in ampoules or in form of commercially available T-2 and HT-2 solutions.

3 Methods

Carry out all procedures at room temperature unless otherwise specified.

3.1 Sample Extraction

1. Weigh 10.0 g of the ground sample (*see* **Note 4**), to the nearest 0.1 g, into a 100 mL flask.
2. Add 50 mL of extraction solvent mixture, and shake for 60 min with an orbital shaker. The motion speed shall be adjusted to obtain the maximum agitation of the extraction mixture (*see* **Note 5**).
3. Filter through a filter paper.
4. Pipette a 5 mL aliquot of the filtered extract into a 10 mL glass tube and evaporate to dryness by the heating block under a gentle stream of air (or nitrogen) at approximately 40 °C.
5. Reconstitute the residue by adding 0.3 mL methanol, vortexing for 1 min, then adding 0.7 mL water and vortexing again for 1 min.

3.2 Extract Clean Up and Preparation of Test Solution

1. Connect the SPE column to the vacuum manifold.
2. Activate the column by passing 2 mL methanol at about 2–3 drops per second, then equilibrate the column by passing 2 mL of conditioning solution at about 2–3 drops per second, avoiding to dry the column after sorbent equilibration.
3. Pass through the column the whole volume (1 mL) of reconstituted extract at a flow rate of about one drop per second and discard the eluate. Dry the column (*see* **Note 6**).
4. Wash the column by passing 1 mL of washing solution at about 1–2 drops per second and discard the eluate. Dry the column (*see* **Note 6**).
5. Elute the toxins in a 4 mL vial with 1 mL methanol at about one drop per second. Dry the column to completely recover the eluate (*see* **Note 6**).
6. Evaporate the cleaned up extract to dryness by the heating block under nitrogen or air stream, at approximately 40 °C.
7. Add 1 mL of injection solvent to the dried residue, cap vial, and mix on vortex mixer for 30 s.
8. Filter the reconstituted extract through a syringe filter and transfer the filtrate (test solution) in an autosampler vial.

3.3 Preparation of Calibrant Solutions

Prepare the matrix-assisted calibration curve for each test material (e.g., oats, oat flakes, *see* **Note 7**).

1. Prepare a mixed mycotoxin solution in acetonitrile containing 1 μg/mL of T-2 toxin and 1 μg/mL of HT-2 toxin (*see* **Note 8**).

Table 1
Preparation of calibrant solutions: volumes of mixed standard solution containing 1 μg/mL of T-2 toxin and 1 μg/mL of HT-2 toxin to be added to the methanol eluate recovered from SPE columns, and resulting T-2 and HT-2 levels

Calibration solution	Mixed standard solution (μL)	T-2/HT-2 level (μg/kg)
1	10	10
2	20	20
3	100	100
4	200	200
5	500	500
6	1000	1000

Table 2
SRM transitions for the determination of T-2 and HT-2 toxins

Analyte	Precursor ion	Q1 (*m/z*)	Q3 (*m/z*)
HT-2	$[HT\text{-}2+NH_4]^+$	442.0	263.3[a] 215.1
T-2	$[T\text{-}2+NH_4]^+$	484.0	215.1[a] 185.3

[a]Quantifier ion

2. Prepare six matrix assisted calibration solutions in six blank sample extracts (*see* **Note 9**) cleaned-up through SPE columns as follow. Add aliquots of T-2 and HT-2 mixed standard solution listed in Table 1 (*see* **Note 10**) to the purified blank extracts (methanol eluate) before drying them down. Dry the spiked eluates (*see* **Note 6**) and reconstitute with 1 mL of LC injection solvent by vortexing for 1 min.

The use of isotope labeled internal standards can be considered as alternative to matrix assisted calibration (*see* **Note 11**).

3.4 Measurements

1. LC conditions: choose an analytical column, mobile phase and gradient settings that let you met the requirements in Subheading 2.3 (*see* **Note 12**).
2. MS conditions: choose an ion source with sufficient ionization yield for the two analytes, and ion source settings such that a stable spray is achieved. Use SRM detection by monitoring two daughter ions for each chosen parent as reported in Table 2 (*see* **Note 13**). Example conditions for suitable LC-MS/MS system and the relevant chromatogram are shown in Fig. 1 (*see* **Note 12**).

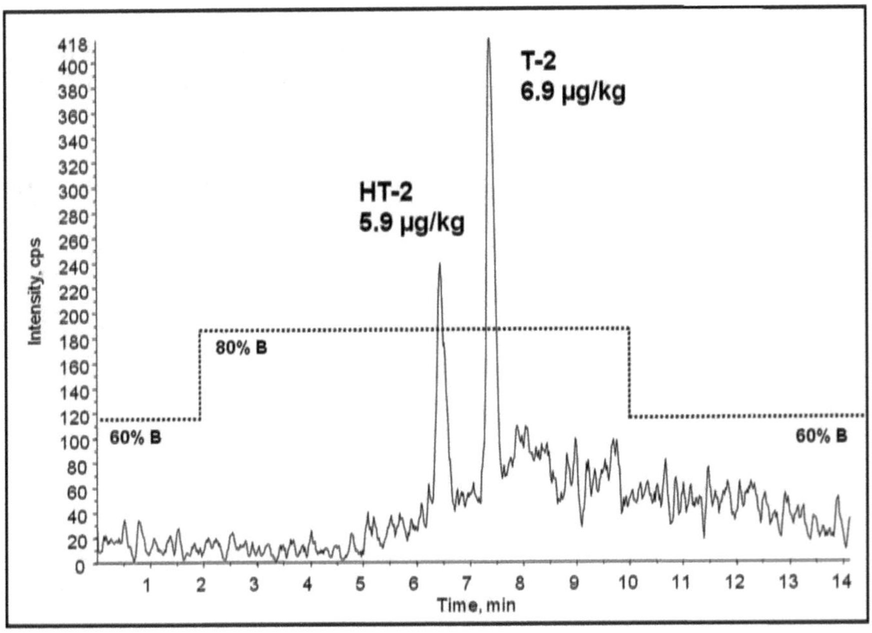

Column: Synergi Hydro (150 x 3mm, 4 µm), Phenomenex
Flow: 400 µL/min
Column oven: 50 °C
Eluent A: H_2O, 5mM $AcNH_4$
Eluent B: CH_3OH, 5mM $AcNH_4$
Injected Volume: 20µL (20 mg sample)

MS Ion source: electrospray (ESI)
Detection mode: Selected Reaction Monitoring (SRM)

Fig. 1 Example of suitable LC-MS/MS conditions for T-2 and HT-2 determination in oats and oat-based products. The total ion chromatogram (sum of SRM transitions) relevant to an oats extract naturally contaminated with T-2 and HT-2 at levels of 6.9 µg/kg (T-2) and 5.9 µg/kg (HT-2) is shown. Chromatographic conditions and main MS conditions are reported in figure; the *dotted line* shows changes in mobile phase composition

3. Inject aliquots of the calibrant solutions and test solutions into the LC-MS/MS apparatus using identical conditions.
4. For each injection measure the peak area of the quantifier ion of each analyte. Build a calibration graph by plotting the signals (peak areas) of the calibration solutions against the corresponding levels (µg/kg) for T-2 and, separately, for HT-2.
5. Estimate slope and intercept of each of the two calibration functions by applying the weighted least-square regression over all data.

6. Using the estimated slopes and intercepts from linear regression calculate the mass fraction of T-2 ($C_{T(T2)}$) and HT-2 ($C_{T(HT2)}$) in micrograms per kilograms in the test solutions as follows:

$$C_{T(T2)} = \frac{\text{signal}_{T2} \quad \text{intercept}_{T2}}{\text{slope}_{T2}} [\ \text{g/kg}]$$

$$C_{T(HT2)} = \frac{\text{signal}_{HT2} \quad \text{intercept}_{HT2}}{\text{slope}_{HT2}} [\ \text{g/kg}]$$

7. In house verification of method performances. Before applying the method to unknown samples, verify method performances in the target commodity by preparing and analyzing a matrix assisted calibration curve, fortified samples (i.e., samples added with a known mycotoxin amount) for recovery evaluation, and/or by analysis of quality control materials (certified reference materials where available) (*see* **Note 14**).

4 Notes

1. T-2 and HT-2 toxins are responsible for immunosuppressive and cytotoxic effects. Studies on dermal exposure have reported T-2 toxin to have extremely toxic effects on skin and mucous surfaces. Observe appropriate safety precautions for handling such compounds. Gloves and safety glasses shall be worn at all times and all standard and sample preparation stages shall be carried out in a fume cupboard.
2. To provide a high level of assurance that the data generated from the LC analysis are reliable, the performance of the LC system should be monitored at regular intervals (e.g., every 6 months). Key performance attributes for a typical LC system include flow rate accuracy, injector precision, and temperature accuracy of the column compartment. Suitable performance attributes for the above mentioned parameters are: flow rate accuracy: ±2 % of the set flow rate; injector precision: 1 % RSD (relative standard deviation of at least six replicate injections); column compartment temperature accuracy: ±2 °C of the set temperature.
3. To ensure reliable analyte identification and quantitative determination, the minimum acceptable retention time for the analytes under examination should be at least twice the retention time corresponding to the void volume of the LC column.
4. Sample preparation and analysis shall be performed with representative and homogeneous laboratory samples. Sampling plays a crucial part in the precision of the determination of the levels of mycotoxins, which are very heterogeneously distributed in a lot. Sampling for mycotoxin control shall be

carried out in accordance with the methods set out in Regulation 401/2006/EC [12]. The laboratory sample shall be finely ground to pass a 500 μm sieve, and mixed thoroughly using a process that has been demonstrated to achieve complete homogenisation.

5. In the extraction step a laboratory blender can be used instead of a shaker. In this case, homogenize the mixture for 3 min at a great speed first.
6. To dry the SPE column after sample loading, washing, and toxin elution steps, carefully pass air through the column by applying a gentle vacuum or by using a syringe.
7. Despite the high sensitivity and selectivity, the influence of matrix components on analyte ionization could limit the accuracy of the LC–MS/MS method. Co-elution of matrix compounds could results in the enhancement or suppression of the analyte chromatographic signal. Therefore matrix-matched calibration is necessary for reliable quantitative analyses. Laboratory experiences have shown that matrix effects are not predictable and may vary from matrix-to-matrix. Therefore no general regression curve can be used for different food commodities, but calibration for each individual food commodity is necessary. Matrix assisted calibrant solutions shall be prepared at the same time as the test samples.
8. To prepare T-2 and HT-2 stock solutions dissolve commercially available crystals or films in acetonitrile. These solutions are stable for at least 2 years if stored at −18 °C.
9. Blank samples are samples containing the target mycotoxins at concentrations below the detection limit of the applied analytical method. Were applicable, the absence of target mycotoxins shall be verified by an official method or a reference method.
10. Once it has been shown that there is linearity, the number of levels listed in Table 1 may be adjusted accordingly to local needs and requirements.
11. For T-2 and HT-2 quantification in the test samples, as alternative to external matrix assisted calibration, the internal standard (ISTD) method can be used, based on fully isotope labeled $^{13}C_{24}$-T2 toxin and $^{13}C_{22}$-HT2 toxin, commercially available as reference solutions in acetonitrile. A suitable procedure is described in the following.
    Prepare, a mixed mycotoxin solution in acetonitrile containing 1 μg/mL of T-2 and 1 μg/mL of HT-2, and a mixed ISTD solution in acetonitrile containing 0.25 μg/mL $^{3}C_{24}$-T2 toxin and 0.25 μg/mL $^{13}C_{22}$-HT2 toxin. When preparing the test sample, add an appropriate volume of $^{13}C_{24}$-T2 toxin and $^{13}C_{22}$-HT2 toxin solutions to the SPE eluate, before drying it down (e.g., the addition of 100 μL of mixed ISDT solution

would result in 25 μg/kg of $^{13}C_{24}$-T2, and 25 μg/kg of $^{13}C_{22}$-HT2 in the test sample). Proceed as described in Subheading 3.2 (**steps 6–8**).

To prepare calibrant solutions add different volumes of the mixed-mycotoxin solution to six glass vials to obtain six calibration levels across the calibration range, then add the same volume of mixed ISTD solution used for the test sample (e.g., 100 μL) to each vial. Proceed as described in Subheading 3.2 (**steps 6–8**).

Inject aliquots of the calibrant solutions and test solutions into the column using identical conditions. For each injection calculate the ratio of the peak area of each analyte to the peak area of the respective labeled analog. Use these ratios to estimate the amount of T-2 and HT-2 in the test sample from the relevant calibration curve.

To prepare calibration curves for T-2 and HT-2 plot the peak area ratios of each analyte calculated in the calibrant solutions against the corresponding amount of analyte injected on column. Estimate slope and intercept of each of the calibration functions by applying the weighted least-square regression over all data.

12. The following LC-MS equipment and settings have proven to give satisfactory results. Chromatographic column: stainless steel (150 × 3 mm i.d.) packed with 4 μm C18 reverse-phase material, f.i. Synergi Hydro (Phenomenex), preceded by a suitable corresponding reverse-phase guard column or guard filter, and conditioned at 50 °C. Injection volume: 20 μL. Flow rate: 0.35 mL/min. Gradient settings as shown in Fig. 1.
    Mass spectrometric detector: triple quadrupole mass spectrometer, equipped with an ESI (atmospheric pressure chemical ionization) interface, f.i. API 5000 from Applied Biosystems. Using the above equipment and settings, detection limits of T-2 and HT-2 in oats and oats based breakfast cereals were 1 μg/kg and 3 μg/kg, respectively.
13. Different types and modes of mass spectrometric detectors provide different degrees of selectivity, which relates to the confidence in identification. The requirements for identification are specified in the Commission Decision 2002/657/EC [13], and are based on a system of identification points. According to this system, the use of two transitions gives four identification points, corresponding to one precursor (1 point) and two product ions (3.0 points), and fulfills EC requirements (at least three identification points) for compound confirmation.
14. Before applying the method to unknown samples, within-laboratory method validation should be performed to provide evidence that a method is fit for the purpose for which it is to

be used. All definitions and requirements for single-laboratory validation of mycotoxin methods are described in CEN/TR 16059:2010 [14]. Furthermore, method validation shall be supported by regular performance verification during routine analysis, through verification of calibration, recoveries, and analysis of quality control materials (certified reference materials, where available).

Acknowledgements

The research leading to these results has received funding by the UK Food Standards Agency (Project Number C03056).

References

1. Canady RA, Coker RD, Egan SK, Krska R, Olsen M, Resnik S, Schlatter J (2001) Safety evaluation of certain mycotoxins in food-BST. WHO food additives series 47. FAO Food and Nutrition Paper 74. WHO, Geneva. pp. 557–638
2. Schothorst R, van Egmond H (2004) Report from Scoop task 3.2.10. "Collection of occurrence data of Fusarium toxins in food and assessment of dietary intake by the population of EU members states" Sub-task: trichothecenes. Toxicol Lett 153:133–143
3. van der Fels-Klerx HJ, Stratakou I (2010) T-2 toxin and HT-2 toxin in grain and grain-based commodities in Europe: occurrence, factors affecting occurrence, co-occurrence and toxicological effects. World Mycotoxin J 3:349–367
4. van der Fels-Klerx HJ (2010) Scientific/technical report—occurrence data of trichotecene mycotoxins T-2 toxin and HT-2 toxin in food and feed. http://www.efsa.europa.eu/en/supporting/doc/66e.pdf. Accessed 28 Jan 2016
5. Schuhmacher-Wolz U, Heine K, Schneider K (2010) Scientific report—report on toxicity data on trichothecene mycotoxins HT-2 and T-2 toxins. http://www.efsa.europa.eu/en/supporting/doc/65e.pdf. Accessed 28 Jan 2016
6. Opinion of the Scientific Committee on Food on *Fusarium toxins*, part 5: T-2 toxin and HT-2 toxin. http://ec.europa.eu/food/safety/docs/cs_contaminants_catalogue_out88_en.pdf. Accessed 28 Jan 2016
7. Edwards SG (2009) *Fusarium* mycotoxin content of UK organic and conventional wheat. Food Addit Contam A 26:496–506
8. Edwards SG (2009) *Fusarium* mycotoxin content in UK organic and conventional oats. Food Addit Contam A 26:1063–1069
9. Commission Recommendation of 27 March 2013 on the presence of T-2 and HT-2 toxin in cereals and cereal products. Off J Eur Union L 91/12
10. Lattanzio VMT, Pascale M, Visconti A (2009) Current analytical methods for trichothecene mycotoxins in cereals. Trends Anal Chem 28:758–768
11. Lattanzio VMT, Solfrizzo M, Visconti A (2008) Determination of trichothecenes in cereals and cereal-based products by liquid chromatography–tandem mass spectrometry. Food Addit Contam A 25:320–330
12. Commission Regulation (EC) No 401/2006 of 23 February 2006 laying down the methods of sampling and analysis for the official control of the levels of mycotoxins in foodstuffs. Off J Eur Union L 70/12
13. Commission Decisiom (EC) No 657/2002 of 12 August 2002 implementing Council Directive 96/23/EC concerning the performance of analytical methods and the interpretation of the results. Off J Eur Communities L 221/8
14. CEN Technical Report CEN/TR 16059:2010. Food analysis—Performance criteria for single laboratory validated methods of analysis for the determination of mycotoxins

Chapter 10

Multiplex Dipstick Immunoassay for Semiquantitative Determination of *Fusarium* Mycotoxins in Oat

Veronica M.T. Lattanzio and Noan Nivarlet

Abstract

A multiplex competitive antibody-based assay in a dipstick format for the simultaneous detection of major of *Fusarium* mycotoxins in oats is described. Ground oats sample is extracted with a mixture of methanol and water. Sample extract is diluted and directly analyzed by a multiplex dipstick. The test kit employs a microwell of reagents containing antibodies linked to gold particles and a dipstick made up of a nitrocellulose membrane were specific capture lines are located. In the presence of oats extract each antibody binds the corresponding mycotoxin before starting to run vertically on the dipstick in the direction of the capture lines. In 10 min red lines rise from the background on the dipstick. Results are interpreted by an optical reader measuring the ratio between each test line and a control line located on the top of the strip.

Key words *Fusarium* toxins, Deoxynivalenol, T-2 toxin, HT-2 toxin, Zearalenone, Dipstick immunoassay, Oats

1 Introduction

Dipsticks are easy-to-use low-cost lateral flow devices (LFDs) widely used for screening purposes to verify compliance with food and feed legislation requirements. Dipsticks involve immunoassays in which the test sample flows by capillary forces along an analytical membrane that contains immobilized immunoreagents. They are generally designed to detect the presence of a substance or a class of substances at the level of interest allowing rapid decision-making. Portable photometric strip readers, designed to read results electronically from commercial LFDs, are available which help evaluate the result of analysis. The increasing relevance of this technology over more consolidated screening techniques has been emphasized in recent reviews [1–5].

When applying the screening method to unknown samples the concept foresees that samples with a positive result of analysis are subjected to confirmatory analysis, whereas samples with a negative result of analysis are considered as compliant and not further

Sebastian Gasparis (ed.), *Oat: Methods and Protocols*, Methods in Molecular Biology, vol. 1536, DOI 10.1007/978-1-4939-6682-0_10, © Springer Science+Business Media LLC 2017 analyzed. Important criteria of screening tests are on the one hand sufficient *sensitivity* to ensure a low rate of false negative results and on the other hand a sufficient *specificity* to avoid too many false positive results. By analogy with chromatographic methods, the new trend is the development of multi-mycotoxin screening methods [6].

Within the field of mycotoxin analysis, besides test strips devoted to mycotoxins that have been regulated for decades such as aflatoxins, immunodiagnostic test strips are also attracting interest in *Fusarium* toxin analysis [5, 7, 8] mainly driven by recent EU regulations [9, 10].

The major *Fusarium* toxins found in cereals and cereal-based products which can be harmful to both human and animal health are deoxynivalenol (DON), T-2 toxin (T-2), HT-2 toxin (HT-2), zearalenone (ZEA) and the fumonisins (FB_1, FB_2). In order to protect human health from exposure to these mycotoxins the European Commission has recently established regulatory limits for DON, ZEA and sum of fumonisins B_1 and B_2 in cereals and cereal-based foods and feeds, while indicative levels of the sum T-2 and HT-2 have been recently issued [9, 10].

The *Fusarium* toxins T-2, HT-2, and DON are the most dominant mycotoxins in oats [11–13]. For unprocessed oat the European Union has established maximum permitted levels of 1.750 μg/kg for DON, 100 μg/kg for ZEA [9], while a maximum indicative level of 1000 μg/kg has been recently set for the sum of T-2 + HT-2 [10].

A prototype multiplex dipstick for the simultaneous determination of major *Fusarium* toxins, namely DON, ZEA, sum of FB_1/FB_2 and sum of T-2/HT-2 (*see* **Note 1**), has been recently developed under the remit of CON*ff*IDENCE, an EU funded project (FP7) aimed at the development of rapid methods for the screening of multiple contaminants in foods and feed [7]. The present work introduces a rapid semiquantitative method for the determination of *Fusarium* mycotoxins in oats, based on the use of the above mentioned multiplex dipstick that is currently available as commercial kit.

2 Materials

2.1 Sample Extraction Components

1. Methanol (analytical grade, 99% purity).
2. Extraction solvent: mix 80 part per volume of methanol and 20 part per volume of water.
3. Laboratory balance, capable to measure 10 mg.
4. High speed blender, with suitable blender jar (250 mL).
5. Calibrated microliter syringe(s) or microliter pipette(s) (100–1000 μL).

2.2 Dipstick Immunoassay Components

1. Dipstick kit, 4-MYCOSENSOR (Unisensor, Liège, Belgium). The test kit comprises the following components (*see* scheme in Fig. 1) (*see* **Note 2,3**):
   (a) A microwell containing a predetermined amount of antibodies linked to gold nanoparticles.
   (b) A dipstick made up of a set of membranes with specific capture lines.
   (c) A migration buffer (4-MYCOBUFFER, whose composition is covered by industrial confidentiality requirements).
2. 8- or 12-channel micropipette (200 μL).
3. Optical reader for dipstick reading and results interpretation (READSENSOR, Unisensor, Liège, Belgium).

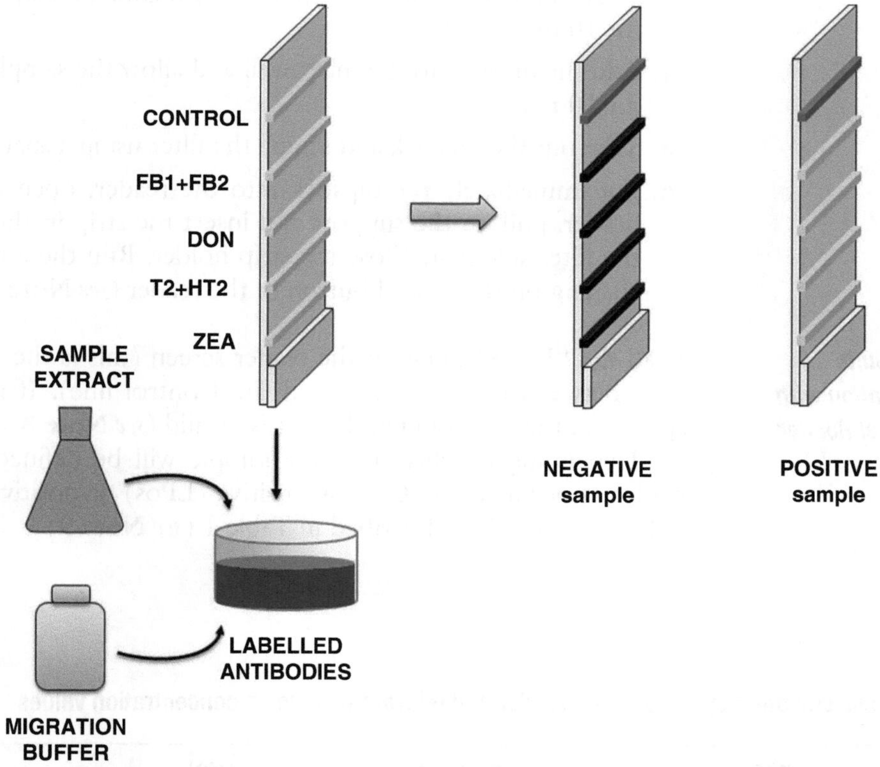

Fig. 1 Scheme of the multiplex dipstick kit. Four test lines (corresponding to ZEA, T-2/HT-2, DON, and FB_1/FB_2 detection lines) and one control line are located on a nitrocellulose membrane. Gold labeled antibodies for ZEA, DON, T-2/HT-2, FB_1/FB_2 and the control line are freeze-dried in the microwells. Intensity of color development at the test lines and at the control line are measured by an optical reader

3 Methods

Carry out all procedures at room temperature unless otherwise specified.

3.1 Sample Preparation

1. Weigh 10 g of oats into a 250 mL blender jar (*see* **Note 4**).
2. Add 50 mL of extraction solvent and extract by blending at high speed for 2 min (*see* **Note 5,6**).
3. Remove the blender cap and allow the sample suspension to settle for 1 min.
4. Pipette 1900 µL of migration buffer in an eppendorf tube and add 100 µL of sample extract. Mix gently by hand.

3.2 Dipstick Analysis

1. Pipette 200 µL of diluted sample extract into the microwell.
2. Mix and homogenize the sample with the freeze-dried reagents into the microwell, by pipetting 15 times. Mix gently to avoid foam formation.
3. Place the microwell into a microwell incubator and incubate for 10 min at 40 °C.
4. Add the dipstick to the microwell and allow the sample to run for 10 min.
5. Take out the dipstick and scrape the filter using a spatula.
6. Put immediately the dipstick into the reader: open the strip holder, pull up the support and insert the strip in the reader, the filter side first. Close the strip holder. Run the reading by clicking on the central button of the reader (*see* **Note 7**).

3.3 Results Interpretation with the Optical Reader

Read the "Ratio" value on the reader screen (this is the intensity ratio between each Test line and the Control line). If no color appears at the control line, the test is invalid (*see* **Note 8**).

Depending on this ratio, the sample will be defined by the reader as negative (NEG), low positive (LPos) or positive (POS) following the values described in Table 1 (*see* **Note 9**).

Table 1
Samples classification by the dipstick reader and relevant mycotoxin concentration values

	ZEA			T-2+HT-2			DON		
Test result	NEG	LPos	POS	NEG	LPos	POS	NEG	LPos	POS
µg/kg	<80	80	>80	<100	100	>100	<1400	1400	>1400

4 Notes

1. Safety measures
   T-2 and HT-2 toxins are responsible for immunosuppressive and cytotoxic effects. Studies on dermal exposure have reported T-2 toxin to have extremely toxic effects on skin and mucous surfaces. The IARC has classified T-2 as belonging to Group 3, i.e., not classifiable as to their carcinogenicity to humans. Methanol is hazardous and flammable. Observe appropriate safety precautions for handling such compounds. Gloves and safety glasses shall be worn at all times and all standard and sample preparation stages shall be carried out in a fume cupboard.
2. Dipstick kits, including strips, reagents microwells, and migration buffer, shall be stored at 4 °C until use. In these conditions kits are stable for 12 months.
3. All kit components, i.e., dipsticks, reagents, microwells, and migration buffer, shall be brought to room temperature just before use.
4. Analysis should be performed with representative and homogeneous laboratory samples. Sampling plays a crucial part in the precision of the determination of the levels of mycotoxins, which are very heterogeneously distributed in a lot. Sampling for mycotoxin control shall be carried out in accordance with the methods set out in Regulation 401/2006/EC [14]. The laboratory sample shall be finely ground to pass a 500 μm sieve.
5. Oat extracts shall be analyzed by dipstick no later than 30 min after the extraction.
6. In the extraction step, the high speed blender can be replaced by the vortex. The vortex extraction shall be carried out for 2 min at the maximum vortex speed, using a suitable vortex tube (f.i. a 50 mL falcon tube).
7. Since the color development at the test line and control line, as well as the color background on the dipstick membrane increases also after removing the pad, for reliable analysis it is recommended to run no more than eight dipsticks at the same time.
8. The color development at the control line ensures the validity of the test. If no color is observed at the control line, the text result is invalid. In these cases the analysis shall be repeated with another dipstick.
9. The extraction protocol described does not guarantee satisfactory recoveries of fumonisins. Fumonisins are not considered in the presented method since these mycotoxins are not likely to contaminate oats.

Acknowledgements

The research leading to these results has received funding from the European Commission's Seventh Framework Programme (FP7/2007–2013) under grant agreement no. 211326.

References

1. Anfossi L, Baggiani C, Giovannoli C, D'Arco G, Giraudi G (2013) Lateral-flow immunoassays for mycotoxins and phycotoxins: a review. Anal Bioanal Chem 405:467–480
2. Krska R, Molinelli A (2009) Rapid test strips for analysis of mycotoxins in food and feed. Anal Bioanal Chem 393:67–71
3. Maragos CM, Busman M (2010) Rapid and advanced tools for mycotoxin analysis: a review. Food Addit Contam A 27:688–700
4. Dzantiev BB, Byzova NA, Ursov AE, Zherdev AV (2014) Immunochromatographic methods in food analysis. Trend Anal Chem 55:81–93
5. Li W, Powers S, Dai SY (2014) Using commercial immunoassay kits for mycotoxins: 'joys and sorrows'? World Mycotoxin J 7:417–430
6. Goryacheva IY, De Saeger S, Eremin SA, Van Peteghem C (2007) Immunochemical methods for rapid mycotoxin detection: evolution from single to multiple analyte screening: a review. Food Addit Contam 24:1169–1183
7. Lattanzio VMT, Nivarlet N, Lippolis V, Della Gatta S, Huet AC, Delahaut P, Granier B, Visconti A (2012) Multiplex dipstick immunoassay for semi-quantitative determination of Fusarium mycotoxins in cereals. Anal Chim Acta 718:99–108
8. Lattanzio VMT, von Holst C, Visconti A (2013) Experimental design for in-house validation of a screening immunoassay kit. The case of a multiplex dipstick for *Fusarium* mycotoxins in cereals. Anal Bioanal Chem 405:7773–7782
9. Commission Regulation (EC) No 1881/2006 of 19 December 2006 setting maximum levels for certain contaminants in foodstuffs. Off J Eur Union L 364/5
10. Commission Recommendation of 27 March 2013 on the presence of T-2 and HT-2 toxin in cereals and cereal products. Off J Eur Union L 91/12
11. Gareis M, Schothorst RC, Vidne A, Bergsten C, Paulsen B, Brera C, Miraglia M (2003) Collection of occurrence data of *Fusarium* toxins in food and assessment of dietary intake by the population of EU member states. Report of Experts Participating in SCOOP Task 3.2.10. http://europa.eu.int/comm/food/fs/scoop/task3210.pdf. Accessed 26 Jan 2016
12. Langseth W, Rundberget T (1999) The occurrence of HT-2 toxin and other trichothecenes in Norwegian cereals. Mycopathologia 147:157–165
13. Scudamore KA, Baillie H, Patel S, Edwards SG (2007) Occurrence and fate of *Fusarium* mycotoxins during commercial processing of oats in the UK. Food Addit Contam 24:1374–1385
14. Commission Regulation (EC) No 401/2006 of 23 February 2006 laying down the methods of sampling and analysis for the official control of the levels of mycotoxins in foodstuffs. Off J Eur Union L 70/12

Chapter 11

Microarray-Based Immunoassay for Parallel Quantification of Multiple Mycotoxins in Oat

Susanna Oswald, Richard Dietrich, Erwin Märtlbauer, Reinhard Niessner, and Dietmar Knopp

Abstract

Because multianalyte methods are highly desirable in order to keep analysis time and costs low, the biosensor development increasingly focuses on parallel analysis of several mycotoxins. Here, we describe an indirect competitive immunoassay on regenerable, reusable glass microchips for the parallel determination of aflatoxins, ochratoxin A, deoxynivalenol, and fumonisin B1 in oat extracts, using a fully automated flow-through device with chemiluminescence readout.

Key words Automated multianalyte immunoassay, Munich chip reader (MCR 3), Chemiluminescence microarray readout, Mycotoxins, Oat

1 Introduction

The mycotoxins are secondary metabolites from fungal species such as *Aspergillus*, *Fusarium*, *Alternaria*, and *Penicillium* growing on agricultural commodities in the field or during storage [1]. They are known for their acute toxic, immunosuppressive, mutagenic, teratogenic, or even carcinogenic effects and therefore, maximum levels for several mycotoxins in food have been set by the European Community (EC) [2, 3]. For example, maximum levels of 4 μg/kg for the sum of aflatoxins B1, B2, G1, and G2 as well as 2 μg/kg for aflatoxin B1 and 5 μg/kg for ochratoxin A in cereals and all products derived from cereals are allowed by the Commission regulations (EC) No 1881/2006 and No 1126/2007. In contrast, higher limits for fumonisin B1 of 400–4000 μg/kg and for deoxynivalenol of 750–1750 μg/kg were set.

Mycotoxins constitute a very heterogeneous group of compounds that differ greatly in their chemico-physical properties and polarity, as well as in their distribution and concentration in food (Fig. 1). Because of this, particular extraction, cleanup, and detection

Sebastian Gasparis (ed.), *Oat: Methods and Protocols*, Methods in Molecular Biology, vol. 1536,
DOI 10.1007/978-1-4939-6682-0_11,

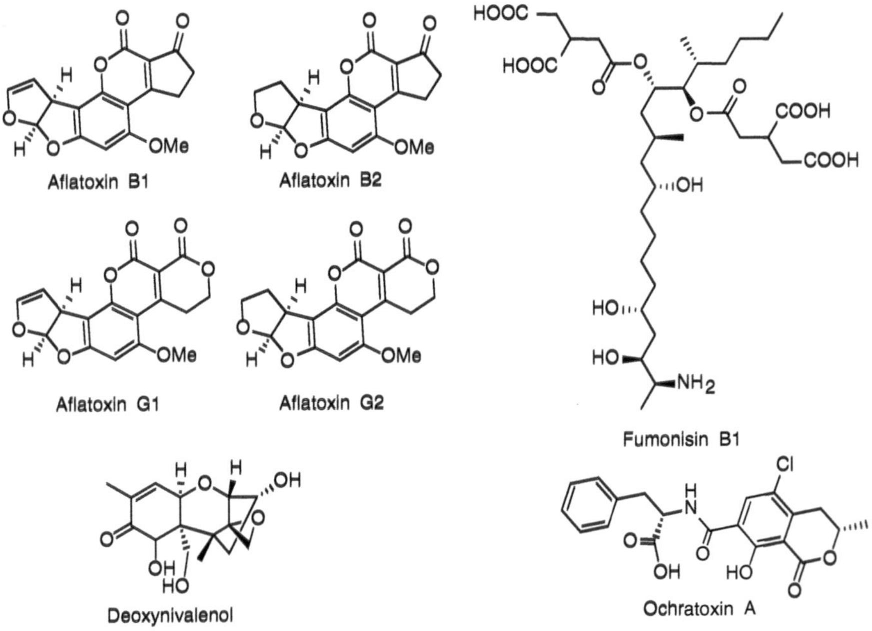

Fig. 1 The chemical structures of aflatoxins B1, B2, G1, G2, deoxynivalenol, fumonisin B1 and ochratoxin A

strategies have been developed. For the qualitative and quantitative analysis of mycotoxins in food traditionally high performance liquid chromatography (HPLC), gas chromatography (GC), and thin-layer chromatography (TLC) are used ([4–6]; and references therein). Over the last decade, HPLC hyphenated to tandem mass spectrometry (LC-MS/MS) have become the method of choice owing to the development of efficient electrospray (ESI) and atmospheric pressure chemical ionization (APCI) interfaces [7–10]. However, an entire sample preparation and especially extract cleanup is still required using, e.g., solid phase extraction (SPE) based on conventional reversed-phase materials, mixed-mode cartridges (MycoSep®), molecularly imprinted polymers (MIP), and immunoaffinity columns (IAC) [11, 12]. Even so, reliable quantification can often be achieved only by matrix-matched calibration or use of isotope-labeled internal standards [5, 11, 13].

To provide rapid and sensitive detection based on cost-effective and easy-to-use methods, which can be performed in a non-laboratory environment by non-specialists, a multitude of the so-called screening methods were developed. Generally, they do not require sample preparation other than extraction and dilution; however, they are less precise and sensitive than chromatographic methods. This field is dominated by immunochemical tests that use anti-mycotoxin antibodies in a variety of different assay formats like, e.g., enzyme-linked immunosorbent assay (ELISA), fluorescence polarisation immunoassay (FPIA), and kinetic FPIA, also called stopped-flow FPIA (*aokinmycontrol*® kits, Aokin AG, Berlin, Germany) [4, 6, 14]. Notably, the need to provide even more simple, fast, robust, user-friendly, and cost-effective methods has led to a multitude of noninstrumental membrane-based lateral flow devices and gel-based flow-through tests with steadily improved detection capability and use of different labels ([15–23]; and references therein). Normally, these tests provide a "yes/no" result for a certain limit or range set previously; however, during the last years there is a tendency to develop modified devices enabling quantitative evaluation and multiplexing.

Because multi-mycotoxin methods are highly desirable in order to keep analysis time and costs low, the biosensor development increasingly focuses on parallel analysis of several mycotoxins, e.g., the MultiAnalyte Profiling (xMAP) technology from Luminex Corp. (Austin, TX) which combines a unique color-coded microsphere suspension array with a dedicated flow cytometer [24–27]. Here, we demonstrate an indirect competitive immunoassay on regenerable, reusable glass microchips for the parallel determination of four mycotoxins, i.e., aflatoxins, ochratoxin A, deoxynivalenol, and fumonisin B1 in cereal extracts on a fully automated flow-through device (MCR 3) with chemiluminescence readout (Fig. 1) [28].

2 Materials

1. Prepare all solutions using ultrapure water (prepared by reverse osmosis with UV treatment to attain a resistivity of 18 MΩ cm at 25 °C) and analytical grade reagents. Prepare and store all reagents at room temperature (RT) unless indicated otherwise. Diligently follow all waste disposal regulations when disposing waste materials. Mycotoxins are highly toxic. Especially, the aflatoxins and ochratoxin A should be handled with extreme care. Related contaminated material should be decontaminated with an aqueous solution of sodium hypochlorite (10%).

2.1 Microarray Fabrication Components, Primary Anti-Mycotoxin Antibodies, and Mycotoxin Standard Solutions

1. Stand alone platform "Munich Chip Reader 3" (MCR 3) (GWK Präzisionstechnik GmbH, München, Germany).
2. Ochratoxin A (OTA) (Sigma-Aldrich, Steinheim, Germany). To prepare stock solution (100 μg/mL), add ~5 mL of acetonitrile to a 10-mL conical flask. Weigh 1.00 mg of OTA and transfer to the flask. Mix and make up to 10 mL with acetonitrile. Prepare 1:100 dilution (v/v) with methanol by the addition of 2 mL of methanol to a 5-mL conical flask followed by 50 μL of OTA stock solution. Mix, make up to 5 mL with methanol, and store at 4 °C.

3. Aflatoxins B1 and B2 (AFB1 and AFB2) (Sigma-Aldrich) (*see* **Note 1**). Prepare stock solution in acetonitrile (100 μg/mL) by a validated method [28]. Add ~5 mL of acetonitrile to a 10-mL CERTAN® capillary bottle (Sigma-Aldrich) (*see* **Note 2**). Weigh about ~1.00 mg of aflatoxin and transfer to the bottle. Mix and make up to 10 mL with acetonitrile. Prepare 1:10 dilution with acetonitrile (v/v) using the same type of capillary bottle. To determine the concentration of aflatoxin, prepare further dilutions in the range of 1:5 to 1:15 (v/v) with acetonitrile and measure absorbance of aliquots (threefold determination) at the absorbance maximum between 355 and 360 nm in UV-transparent plastic cuvettes. The optimal absorbance should be 0.6 ± 0.2. Calculate the AFB1 and AFB2 concentrations using the validated molar extinction coefficients (ε) in acetonitrile of 20700 and 22500, respectively [29]. Store stock solution and dilutions at 4 °C.
4. Deoxynivalenol (DON) (Sigma-Aldrich). To prepare stock solution (1 mg/mL) add ~5 mL of acetonitrile to a 10-mL conical flask. Weigh 10.00 mg of DON and transfer to the flask. Mix and make up to 10 mL with acetonitrile. Prepare 1:10 dilution (v/v) with methanol by the addition of 2 mL of methanol to a 5-mL conical flask followed by 500 μL of DON stock solution. Mix, make up to 5 mL with methanol, and store at 4 °C.
5. HS-DON (3,15-*O*-dihemisuccinyl-DON) is prepared by the reaction of DON with succinic anhydride. 10 mg of DON is dissolved with 200 mg of succinic anhydride in 1 mL of pyridine and heated at 100 °C in a steam bath for 8 h. Pyridine is removed under N_2 pressure and the residue dissolved in 15 mL of methylene chloride. After extraction three times with 10 mL of 0.1 M HCl the organic phase is dried under vacuum in a rotary evaporator and the residue redissolved in 2 mL of methanol (*see* **Note 3**).
6. Fumonisin B1 (FB1) (Sigma-Aldrich). To prepare stock solution (150 μg/mL), add ~3 mL of methanol to a 10-mL conical flask. Weigh 1.50 mg of FB1 and transfer to the flask. Mix, make up to 10 mL with methanol, and store at 4 °C.
7. Mouse monoclonal anti-ochratoxin (OTA) antibody 5G9 (concentration of 1.0 mg/mL); (Soft Flow Biotechnology, Gödöllö, Hungary).
8. Mouse monoclonal anti-Aflatoxin antibody 1F2 (concentration 1.0 mg/mL).
9. Mouse monoclonal anti-Deoxynivalenol antibody 1H8 (concentration 0.72 mg/mL).
10. Mouse monoclonal anti-Fumonisin B1 antibody 1H4 (concentration 0.37 mg/mL).

11. Double-sided adhesive foil ARcare 90106 (Adhesive Research Ireland Ltd., Limerick, Ireland).
12. Carbonate buffer: Add about 80 mL water to a 100-mL conical flask. Weigh 0.159 g sodium carbonate and 0.29 g sodium bicarbonate and transfer to the flask. Mix and adjust pH to 8.5 with HCl. Make up to 100 mL with water. Store at 4 °C.
13. Spotting buffer: Transfer 300 μL dimethylsulfoxide (DMSO), 690 μL carbonate buffer, and 10 μL glycerol to a 1-mL conical flask. Mix and store at 4 °C.
14. BioOdyssey Calligrapher Miniarrayer (Bio-Rad Laboratories GmbH, München, Germany).
15. Solid printhead for microcontact spotting (Stealth printhead SNS12; Arrayit Corporation, Sunnyvale, USA).
16. T25 Ultra-Turrax® dispenser (IKA Labortechnik, Staufen, Germany).

2.2 Automated Chemiluminescence Immunoassay Components

1. Phosphate buffer saline (PBS, pH 7.6): Add about 100 mL water to a 1-L conical flask. Weigh 1.36 g potassium dihydrogen phosphate, 12.2 g dipotassium hydrogen phosphate, and 8.5 g sodium chloride and transfer to the flask. Make up to 1 L with water. Store at 4 °C.
2. Running buffer: PBS containing 0.5 % (w/v) casein. Add about 90 mL PBS to a 100-mL conical flask. Weigh 0.5 mg casein and transfer to the flask. Make up to 100 mL with water (*see* **Note 4**). Store at 4 °C.
3. Regeneration buffer: Add about 1.5 L water to a 2-L conical flask. Weigh 15.01 g glycine, 11.7 g sodium chloride, and 2 g sodium dodecyl sulphate (SDS) and transfer to the flask. Mix and adjust pH to 3.0 with HCl. Make up to 2 L with water. Store at 4 °C.
4. Wash buffer: Add 20 mL methanol to a 100-mL conical flask and make up to 100 mL with PBS. Prepare fresh and use on the same day.
5. Primary antibodies mixture (Ab-mix): Prepare mixture in running buffer. Add aliquots of primary antibody stock solutions to 30 mL of running buffer to obtain dilutions of 1:15,000 of anti-aflatoxin antibody, 1:5000 of anti-OTA antibody, 1:4000 of anti-FB1 antibody, and 1:3000 of anti-DON antibody (*see* **Note 5**).
6. Peroxidase-labeled horse anti-mouse IgG: Prepare ~1:1000 dilution in running buffer of by the addition of 50 μL of antibody stock solution to 50 mL of buffer.
7. SuperSignal ELISA Femto Substrate (Thermo Fisher Scientific GmbH, Dreieich, Germany). The kit constitutes of reagent *A* (luminol solution) and reagent *B* (H_2O_2 solution). The reagents are used without further treatment.

3 Methods

Carry out all procedures at RT unless otherwise specified. Before analysis, return refrigerated samples to RT before analysis.

3.1 Fabrication of Functionalized Glass Slides

1. For cleaning, put the glass slides (conventional microscopy glass slides (26 × 72 × 1 mm) into a slide chamber filled with 200 mL water and 4 mL of Hellmanex®III solution (Hellmanex GmbH & Co. KG, Müllheim, Germany) and incubate for 1 h in an ultrasonic bath, followed by slight orbital shaking overnight and repeated ultrasonication for 1 h. Wash the slides five times with 200 mL water and make final wash with methanol. Dry the slides carefully with a stream of nitrogen and perform visual surface inspection.
2. To activate the glass surface (to obtain many hydroxyl groups) incubate the slides with 200 mL methanol/HCl (1:1, v/v) for 1 h under shaking, followed by washing five times with 200 mL water and incubation for another 1 h with concentrated sulphuric acid. Wash sequentially with water and methanol and dry the slides under nitrogen flow.
3. The pretreated glasses are silanized by the addition of 600 μL of 3-glycidyloxypropyltrimethoxysilane (GOPTS) on the chip surface and covered with a second glass slide (sandwich technique). After incubation at RT for 3 h separate the slides from each other in pure ethanol and clean by sonication subsequently with ethanol, methanol, and ethanol each for 15 min.
4. After drying under nitrogen flow and heating at 98–100 °C in a drying oven for 5 min, cover the silanized glass slide with 1 mL molten diamino-poly(ethylene glycol) (DAPEG) (*see* **Note 6**) and incubate at 98–100 °C for 15 h by using the sandwich technique again. Finally, separate the slides and sonicate intensively with water for two times each 15 min and dry under nitrogen flow (*see* **Note 7**). Store the slides in a desiccator under vacuum, protected from light, until further use.

3.2 Microarray Construction

1. Apply the DAPEG-functionalized glass slides directly to the spotting process, using a Miniarrayer (Fig. 2) (*see* **Notes 8** and **9**). Dissolve 2.5 μL of an aqueous solution (10 mg/mL) of aflatoxin B2-carboxymethyl oxime (AFB2-CMO) in 45 μL of spotting buffer (*see* **Notes 10** and **11**). Dissolve 5 μL of OTA (5.5 mg/mL stock in acetonitrile) and 5 μL of FB1 (10 mg/mL stock in methanol), in 45 μL spotting buffer. Dissolve 20 μL of HS-DON (10 mg/mL stock in methanol) in 30 μL of spotting buffer. Use spotting buffer as a negative control.
2. To introduce an activated carboxyl group into AFB2-CMO, OTA, and HS-DON, 10 Eq of *N*-(3-dimethylaminopropyl)-*N*′-ethylcarbodiimide hydrochloride (EDC), 10 Eq

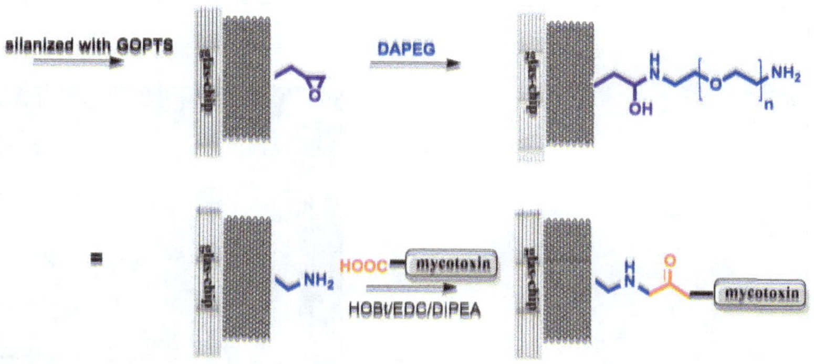

Fig. 2 Immobilization of mycotoxins to the modified glass-chip: After silanization with GOPTS, DAPEG is covalently bond to obtain free terminal amino groups. This is followed by coupling of the mycotoxins via available carboxyl functions

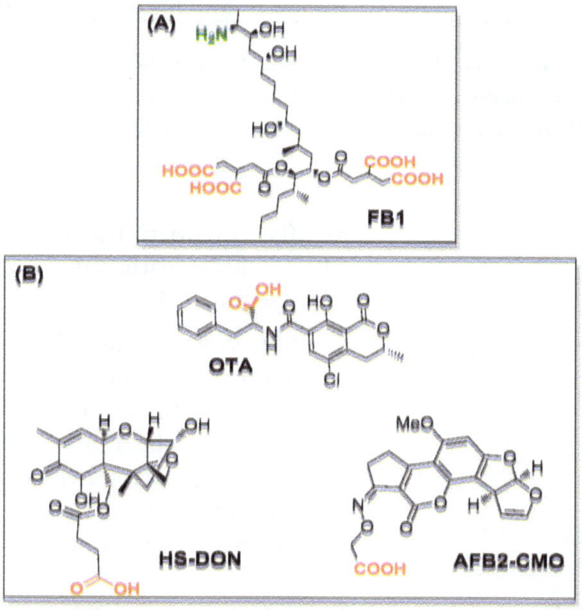

Fig. 3 Structures of immobilized mycotoxins: (**a**) FB1—direct binding via carboxyl groups; (**b**) OTA and derivatized toxins DON and AFB2—binding via activated carboxyl functions

1-hydroxybenzotriazole hydrate (HOBt), and 50 μL *N,N*-diisopropylethylamine (DIPEA) are added to the spotting buffer containing the individual mycotoxin derivative and incubated for 3 h under shaking at RT (Fig. 3) (*see* **Note 12**).

3. Transfer 25 μL of each of the spotting solutions (use plain spotting buffer as the negative control) to different cavities of a 384-well low binding microtiter plate and spot aliquots on the DAPEG glass slide in six replicates. During spotting temper the glass slide at 20 °C and set spotting chamber humidity

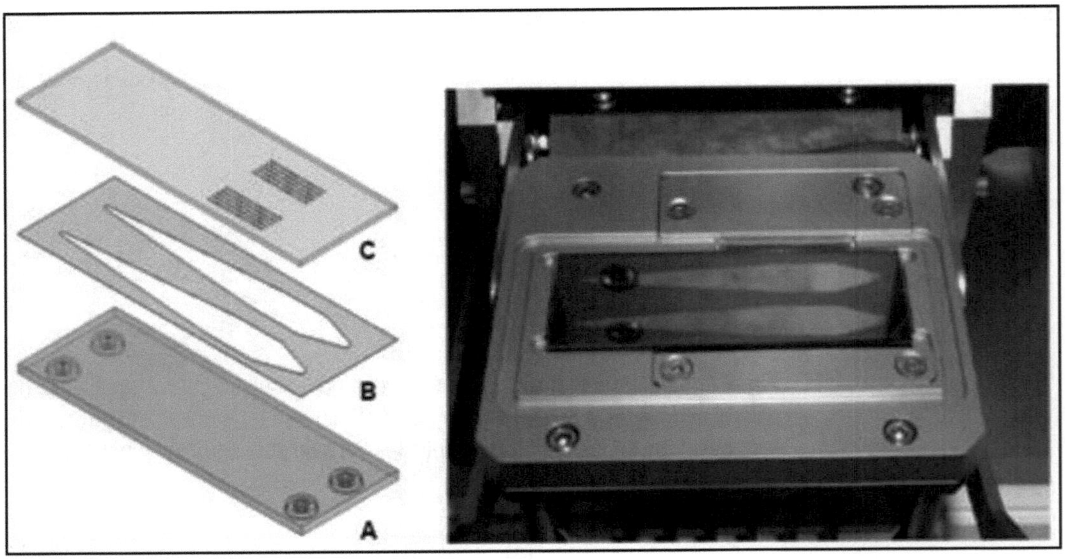

Fig. 4 *Left*: The flow cell construct forming two diamond-shaped chambers, consisting of a black poly(methyl methacrylate) (PMMA) opaque slide that contains two openings at opposite ends for the inlet and the outlet of the flow cell (*A*), double-sided adhesive sheet (*B*), and spotted glass slide (*C*). *Right*: MCR 3 flow cell holder

to 50%. Incubate the slides for 15 h at RT, then wash with PBS, sonicate with water for 15 min, and dry under nitrogen flow (*see* **Note 13**).

4. Glue the slides to a double-sided adhesive sheet where the shape of the flow cell had been previously carved out by laser-cutting. Finish the assembled chip and flow cell by placing a double-sided adhesive sheet in between the spotted glass chip (on top) and a black poly(methyl methacrylate) (PMMA) opaque slide (at the bottom) which contains two openings at opposite ends for the inlet and the outlet of the flow cell (Fig. 4).

3.3 Sample Preparation

1. For extraction, in a beaker mix 25 g of oat sample (*see* **Note 14**) with 5 g of sodium chloride and 100 mL of methanol/water (80:20, v/v) with a dispenser for 5 min while cooling on ice (*see* **Note 15**).
2. Filter the extract through a folded filter (210 mm), and dilute fourfold with PBS (add 20 mL of extract to 60 mL PBS in a 100-mL beaker) to give a final methanol content of 20% (*see* **Note 16**). The diluted samples should be analyzed within 24 h (*see* **Note 17**).

3.4 Performance of the Automated Chemiluminescence Immunoassay (CLISA)

1. Before starting the measurements, pump 3 × 1 mL running buffer with 200 μL/s through the flow cell (Fig. 5) (*see* **Note 18**).
2. The sequence of the performed immunoassay is programmed as follows:

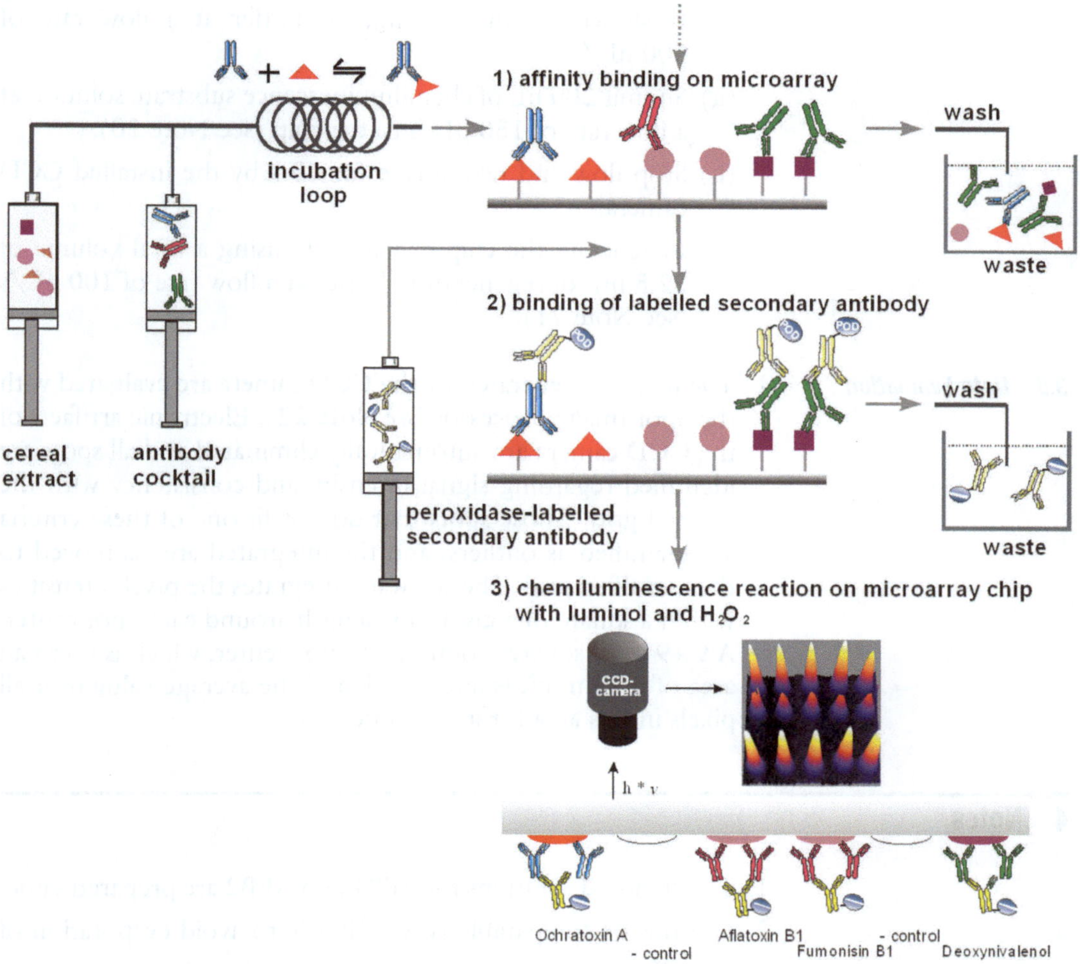

Fig. 5 Principle of the indirect competitive chemiluminescence ELISA using the microarray chip. The mycotoxins are immobilized in a defined grid pattern. After primary and secondary antibodies bound to the chip surface, chemiluminescence reagents (luminol/H_2O_2) are added and the light signal is detected with a CCD camera. The light intensity is inversely proportional to the concentration of mycotoxins in the sample

(a) Submit 1 mL of running buffer to the chip with a flow speed of 500 μL/s.

(b) Pump 500 μL of oat sample and same volume of Ab-mix simultaneously into an incubation loop at a flow rate of 60 μL/s.

(c) Push the sample/antibodies mixture over the chip with 1 mL of running buffer at a flow rate of 100 μL/s.

(d) Wash with 2 mL of running buffer at a flow rate of 500 μL/s.

(e) Submit 1 mL of diluted peroxidase-labeled horse anti-mouse IgG (secondary antibody) to the chip in two steps: 200 μL at a flow rate of 100 μL/s and followed by 800 μL at a flow rate of 10 μL/s (*see* **Note 19**).

(f) Wash with 2 mL of running buffer at a flow rate of 500 μL/s.

(g) Submit 200 μL of chemiluminescence substrate solution at a flow rate of 150 μL/s to the chip (see Note 20).

(h) Stop flow and take image for 60 s by the installed CCD camera.

(i) Regenerate the chip exhaustively using a total volume of 22.5 mL of regeneration buffer at a flow rate of 100 μL/s (see Note 21).

3.5 Data Evaluation

1. The text files generated by the CCD camera are evaluated with the Spot Image Processor (*see* **Note 22**). Electronic artifacts of the CCD camera are automatically eliminated and all spots are identified regarding signal intensity and consistency with the spotted grid. Those spots that do not fit one of these criteria are identified as outliers, and the integrated area is moved to the identified grid. The software integrates the pixel intensities within a square of a given side length around each spot center. A 9 × 9 pixel square around each spot center, which is up to an area of 0.13 mm^2, is integrated, and the average value over all pixels in this area is the spot intensity.

4 Notes

1. The standard solutions for AFB1 and AFB2 are prepared same.
2. Use this or comparable type of bottle to avoid evaporation of solvent.
3. The HS-DON was prepared as described in the literature [30].
4. To improve solubility, heat the running buffer up to 80 °C, filtrate and cool down the filtered solution to RT.
5. In principle, anti-mycotoxin antibodies from other sources (suppliers) can be used providing that the specificity and required stability in appropriately diluted organic solvents acetonitrile and methanol is given. The individual dilutions in running buffer have to be identified.
6. Melt DAPEG in a drying oven at 98–100 °C.
7. The slides can be stored for further use under vacuum but should be used within 1 month.
8. In principal, any spotting machine equipped with heating and humidity adjustment is applicable.
9. The delivered volume is ~5 nL.
10. Use 0.5-mL polypropylene Eppendorf Safe-Lock Tubes, no. 0030121.023, Eppendorf Vertrieb Deutschland GmbH, Wesseling-Berzdorf, Germany.

11. The synthesis of AFB2-carboxymethyl oxime (AFB2-CMO) should be performed as follows. Add 30 mg of *O*-(carboxymethyl)hydroxylamine hemihydrochloride (Sigma-Aldrich) and 10 mg of AFB2 dissolved in 12 mL of methanol/water/pyridine (4:1:1, v/v/v) to a 50-mL round flask and reflux for 3 h at 85 °C. During reaction, the colourless solution shows yellow fluorescence. After reaction, remove the solvent with a rotary evaporator and redissolve the residue with 5 mL of dichloromethane. Wash with 15 mL of 5 % acetic acid that is accompanied by dark-green coloring of the organic phase. After separation of the two phases with a separatory funnel remove the organic solvent with rotary evaporation. Redissolve the black residue with 4 mL acetonitrile (0.1 % trifluoroacetic acid)/water (0.1 % trifluoroacetic acid) (2:1, v/v) and centrifuge for 5 min at 1789 rcf (Hettich Universal Centrifuge 30F, Andreas Hettich GmbH, Tuttlingen, Germany). Dispense the supernatant into two identical aliquots and purify with preparative HPLC using acetonitrile/water as mobile phase and C18 column (*00F-4435-N0* Gemini® 5 μm C18 110 Å, LC Column 150 × 10 mm, Phenomenex, Aschaffenburg, Germany). Start with 10% acetonitrile and increase to 90% within 10 min and keep for further 10 min. Use UV/vis detection at wavelength of 360 nm. Collect fraction between 10 and 14 min and remove solvent with rotary evaporator.
12. Add 7.00 mg HOBt, 8.00 mg EDC and 50 μL DIPEA to 1.00 mL of spotting buffer. After short sonication (15 s) add the specified volumes of mycotoxin solutions. The FB1 contains several carboxyl groups and primary amino group. To avoid intermolecular cross-linking of this mycotoxin, it was immobilized on the chip surface without activation.
13. The slides can be stored for further use under vacuum but should be used within 1 month.
14. Oat, flour, and oat flakes are applicable.
15. Other commercial dispersers are applicable; avoid solvent loss.
16. This corresponds to a 16-fold dilution of the mycotoxin concentration in the original sample.
17. Store samples at 4 °C if not analyzed immediately. Before analysis, return samples to RT.
18. This step is important to displace possibly existing air bubbles.
19. Horseradish peroxidase-labeled anti-mouse IgG antibodies from other suppliers and species (e.g., goat) can be used. The individual dilutions in running buffer have to be identified.
20. Alternately 50-μL aliquots of reagents *A* and *B* are pumped into coil and total volume of 400 μL is pushed over the chip.

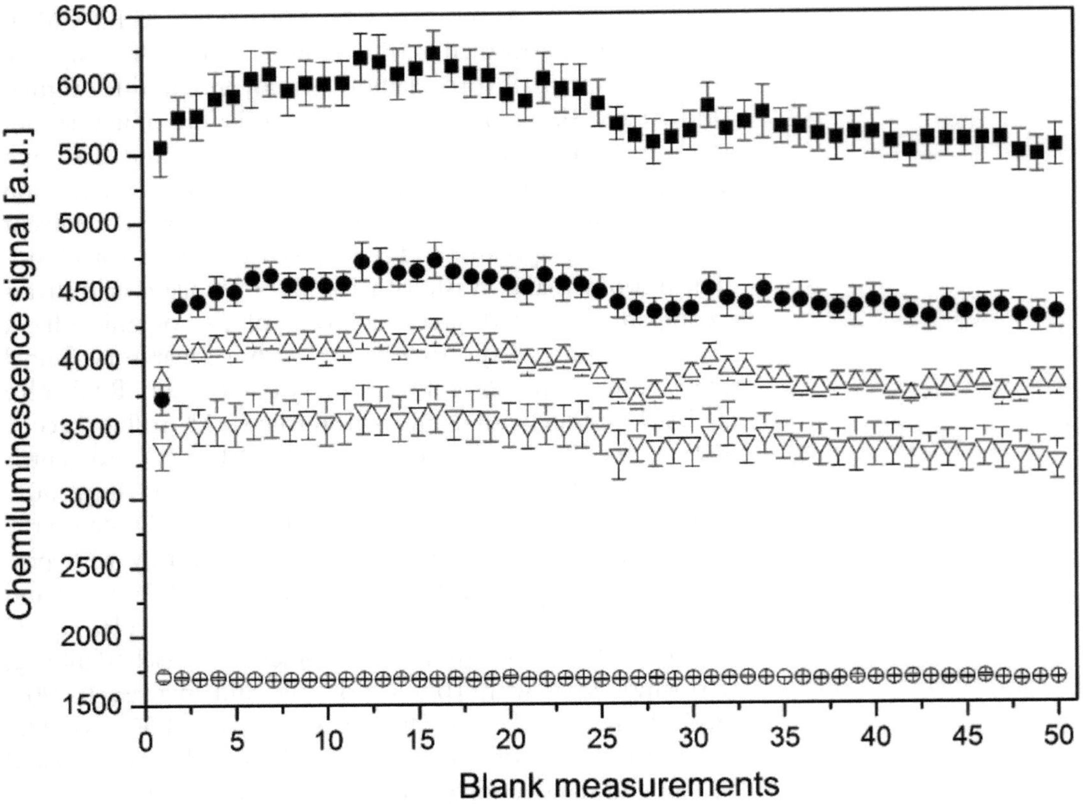

Fig. 6 The performance of 50 regeneration cycles using oat extract as a sample (blank): (*filled square*) aflatoxin B2, (*filled circle*) ochratoxin A, (*triangle*) DON-HS, (*inverted triangle*) fumonisin B1, (*open circle*) negative control. The between-spot standard deviations (1s; $n = 6$) are indicated as error bars

21. On the same day, the chip can be used for at least 50 cycles of sample loading and regeneration (Fig. 6).
22. The text files generated by the CCD camera are evaluated with the Spot Image Processor SIP 0.4 (Karsunke Softwarebüro, Wolnzach, Germany). Appropriate microarray image analysis software from other sources could be applicable.

Acknowledgment

This work was supported by the German Ministry of Economics and Technology (via AiF) and the FEI (Forschungskreis der Ernährungsindustrie e.V. Bonn); project AiF 381 ZN. We are also grateful to Dr. G. Lystik (Soft Flow Biotechnology, Gödöllö, Hungary) for providing the anti-OTA monoclonal antibody and Huntsman Corporation for the free samples of DAPEG.

References

1. Shephard GS (2008) Aflatoxin analysis at the beginning of the twenty-first century. Chem Soc Rev 37:2468–2477
2. Commission regulation (EC) No 1881/2006 Official Journal of the European Union 2006, 364:5–24
3. Commission regulation (EC) No 1126/2007 Official Journal of the European Union 2007, 255:14–17
4. Köppen R, Koch M, Siegel D, Merkel S, Maul R, Nehls I (2010) Determination of mycotoxins in foods: current state of analytical methods and limitations. Appl Microbiol Biotechnol 86:1595–1612
5. Kralj Cigic I, Prosen H (2009) An overview of conventional and emerging analytical methods for the determination of mycotoxins. Int J Mol Sci 10:62–115
6. Shephard GS, Berthiller F, Burdaspal PA, Crews C, Jonker MA, Krska R, MacDonald S, Malone RJ, Maragos C, Sabino M, Solfrizzo M, Van Egmond HP, Whitaker TB (2012) Developments in mycotoxin analysis: an update for 2009–2010. World Mycotoxin J 5:3–30
7. Hajslova J, Zachariasova M, Cajka T (2011) Analysis of multiple mycotoxins in food. In: Zweigenbaum J (ed) Mass spectrometry in food safety: methods and protocols, vol 747, Methods in molecular biology. Humana Press, New York, pp 233–258
8. Herebian D, Zühlke S, Lamshöft M, Spiteller M (2009) Multi-mycotoxin analysis in complex biological matrices using LC-ESI/MS: experimental study using triple stage quadrupole and LTQ-Orbitrap. J Sep Sci 32:939–948
9. Sulyok M, Krska R, Schuhmacher R (2010) Application of an LC-MS/MS based multimycotoxin method for the semi-quantitative determination of mycotoxins occurring in different types of food infected by moulds. Food Chem 119:408–416
10. Monbaliu S, van Poucke C, Detavernier C, Dumoulin F, van de Velde M, Schoeters E, van Dyck S, Averkieva O, van Peteghem C, de Saeger S (2010) Occurrence of mycotoxins in feed as analyzed by a multi-mycotoxin LC-MS/MS method. J Agric Food Chem 58:66–71
11. Krska R, Schubert-Ullrich P, Molinelli A, Sulyok M, MacDonald S, Crews C (2008) Mycotoxin analysis: an update. Food Addit Contam 25:152–163
12. Cichna-Markl M (2011) New strategies in sample clean-up for mycotoxin analysis. World Mycotoxin J 4:203–215
13. Cervino C, Asam S, Knopp D, Rychlik M, Niessner R (2008) Use of isotope-labeled aflatoxins for LC-MS/MS stable isotope dilution analysis in food. J Agric Food Chem 56:1873–1879
14. Li P, Zhang Q, Zhang W (2009) Immunoassay for aflatoxins. Trends Anal Chem 28:1115–1126
15. Liao J, Li H (2010) Lateral flow immunodipstick for visual detection of aflatoxin B_1 in food using immuno-nanoparticles composed of a silver core and a gold shell. Microchim Acta 171:289–295
16. Tang D, Sauceda JC, Lin Z, Ott S, Basova E, Goryacheva I, Biselli S, Lin J, Niessner R, Knopp D (2009) Magnetic nanogold microspheres-based lateral-flow immunodipstick for rapid detection of aflatoxin B2 in food. Biosens Bioelectron 25:514–518
17. Basova EY, Goryacheva IY, Rusanova TY, Burmistrova NA, Dietrich R, Märtlbauer E, Detavernier C, Van Peteghem C, de Saeger S (2010) An immunochemical test for rapid screening of zearalenone and T-2 toxin. Anal Bioanal Chem 397:55–62
18. Beloglazova NV, Speranskaya ES, De Saeger S, Hens Z, Abé S, Goryacheva IY (2012) Quantum dot based rapid tests for zearalenone detection. Anal Bioanal Chem 403:3013–3024
19. Lattanzio VMT, Nivarlet N, Lippolis V, Gatta DS, Huet A-C, Delahaut P, Granier B, Visconti A (2012) Multiplex dipstick immunoassay for semi-quantitative determination of fusarium mycotoxins in cereals. Anal Chim Acta 718:99–108
20. Zheng MZ, Richard JL, Binder J (2006) A review of rapid methods for the analysis of mycotoxins. Mycopathologia 161:261–273
21. Li P, Zhang Z, Zhang Q, Zhang N, Zhang W, Ding X, Li R (2012) Current development of microfluidic immunosensing approaches for mycotoxin detection via capillary electromigration and lateral flow technology. Electrophoresis 33:2253–2265
22. Njumbe Ediage E, di Mavungu JD, Goryacheva IY, van Peteghem C, de Saeger S (2012) Multiplex flow-through immunoassay formats for screening of mycotoxins in a variety of food matrices. Anal Bioanal Chem 403:265–278
23. Anfossi L, Baggiani C, Giovannoli C, D'Arco G, Giraudi G (2013) Lateral-flow immunoassays for mycotoxins and phycotoxins: a review. Anal Bioanal Chem 405:467–480
24. Maragos CM (2009) Biosensors for mycotoxin analysis: recent developments and future prospects. World Mycotoxin J 2:221–238
25. Moises SS, Schäferling M (2009) Toxin immunosensors and sensor array for food quality control. Bioanal Rev 1:73–104

26. Tothill IE (2011) Biosensors and nanomaterials and their application for mycotoxin determination. World Mycotoxin J 4:361–374

27. Peters J, Bienenmann-Ploum M, De Rijk T, Haasnoot W (2011) Development of a multiple flow-cytometric microsphere immunoassay for mycotoxins and evaluation of its application in feed. Mycotoxin Res 27:63–72

28. Oswald S, Karsunke XYZ, Dietrich R, Märtlbauer E, Niessner R, Knopp D (2013) Automated regenerable microarray-based immunoassay for rapid parallel quantification of mycotoxins in cereals. Anal Bioanal Chem 405:6405–6415

29. Nesheim S, Trucksess MW, Page SW (1999) Molar absorptivities of aflatoxins B1, B2, G1, and G2 in acetonitrile, methanol, and toluene-acetonitrile (9 + 1) (modification of AOAC official method 971.22): collaborative study. J AOAC Int 82:251–258

30. Usleber E, Märtlbauer E, Dietrich R, Terplan G (1991) Direct enzyme-linked immunosorbent assays for the detection of the 8-ketotrichothecene mycotoxins deoxynivalenol, 3-acetyldeoxynivalenol, and 15-acetyl deoxynivalenol in buffer solutions. J Agric Food Chem 39:2091–2095

Part IV

Genomic Studies

Chapter 12

M13-Tailed Simple Sequence Repeat (SSR) Markers in Studies of Genetic Diversity and Population Structure of Common Oat Germplasm

Agnieszka Onyśk and Maja Boczkowska

Abstract

Simple Sequence Repeat (SSR) markers are one of the most frequently used molecular markers in studies of crop diversity and population structure. This is due to their uniform distribution in the genome, the high polymorphism, reproducibility, and codominant character. Additional advantages are the possibility of automatic analysis and simple interpretation of the results. The M13 tagged PCR reaction significantly reduces the costs of analysis by the automatic genetic analyzers. Here, we also disclose a short protocol of SSR data analysis.

Key words *Avena sativa* L, Diversity, Germplasm, M13, Population structure, SSR

1 Introduction

Microsatellites are a class of repetitive DNA present in both coding and noncoding regions. They have been identified in the three eukaryote genomes: nuclear, chloroplast, and mitochondrial [1, 2]. The nuclear microsatellites occur frequently in all taxa examined [3, 4]. They consist of di-, tri-, tetra-, penta-, or hexanucleotide tandemly repeated units. The most frequent are dinucleotide repeats, for example $(AT)n$ in plants, whereas the $(AC)n$ repeat is the most common in animals [5–7]. The length of a microsatellite locus is variable and typically ranges from 5 to 40 repeats, but longer tandems are also possible [8]. The variation arises from frequent mutations by slippage and proofreading errors during DNA replication [9]. A microsatellite locus is flanked by unique DNA sequences. These regions are conserved within the species and sometimes among related species, and they are used to design specific primers for polymerase chain reaction (PCR).

SSRs have become a marker of choice for an array of applications in plants due to their hypervariable and codominant nature,

Sebastian Gasparis (ed.), *Oat: Methods and Protocols*, Methods in Molecular Biology, vol. 1536,
DOI 10.1007/978-1-4939-6682-0_12, © Springer Science+Business Media LLC 2017

extensive genome coverage, high reproducibility, and moderate costs and technical demands. Despite numerous advantages the SSRs also have several drawbacks. They are species-specific or poorly transferable among species and have hidden allelic diversity. What is more, primer design requires knowledge of the sequences for the flanking regions.

SSRs are amplified during simple PCR reactions with unlabeled or fluorescently labeled primers. The alleles differ in length so they can be distinguished by classical high resolution electrophoresis in polyacrylamide gels or plate/capillary sequencers. The diversity and population structure studies require that numerous loci be analyzed, and not all of them will be polymorphic within the tested materials. The necessity of fluorescent labeling of one primer of each pair substantially increases the cost of analysis. This drawback can be easily overcome by the M13 tagged PCR reaction [10]. This is a single-tube, nested PCR performed with three primers, i.e., the fluorescent-labeled M13(-21) primer (the universal primer), a specific forward primer fitted with an adapter sequence M13(-21) at its 5′ end, and a specific reverse primer. During the initial cycles of the PCR reaction the M13(-21) tailed forward and reverse primers amplify a microsatellite locus, while in the final stages after using up the scarce forward primer the annealing temperature is lowered to facilitate annealing of the universal fluorescent-labeled M13(-21) primer as the forward primer. Thus, the fluorescent dye is incorporated into PCR products of the microsatellite locus (Fig. 1).

2 Materials

2.1 Plant Material

A set of common oat accessions selected for analysis of the genetic diversity and population structure.

2.2 Molecular Markers

1. Choose from developed SSR markers of oat available in the databases (*see* **Note 1**).
2. Create a forward DNA primer tailed with the M13(-21) sequence: before synthesis, to each forward primer add the M13(-21) sequence to the 5′ end: 5′-CACGACGTTGTAAAACGAC-3′.
3. Create M13 fluorescent dye-labeled DNA primers: synthesize the M13(-21) primers (5′-CACGACGTTGTAAAACGAC-3′) labeled with one of the four fluorescent dyes: FAM, VIC, NED, or PET (*see* **Note 2**).
4. Synthesize a standard length reverse DNA primer.

2.3 DNA Extraction Components

1. DNA extraction kit.
2. “Ultrapure” water (18.2 MΩ cm at 25 °C).

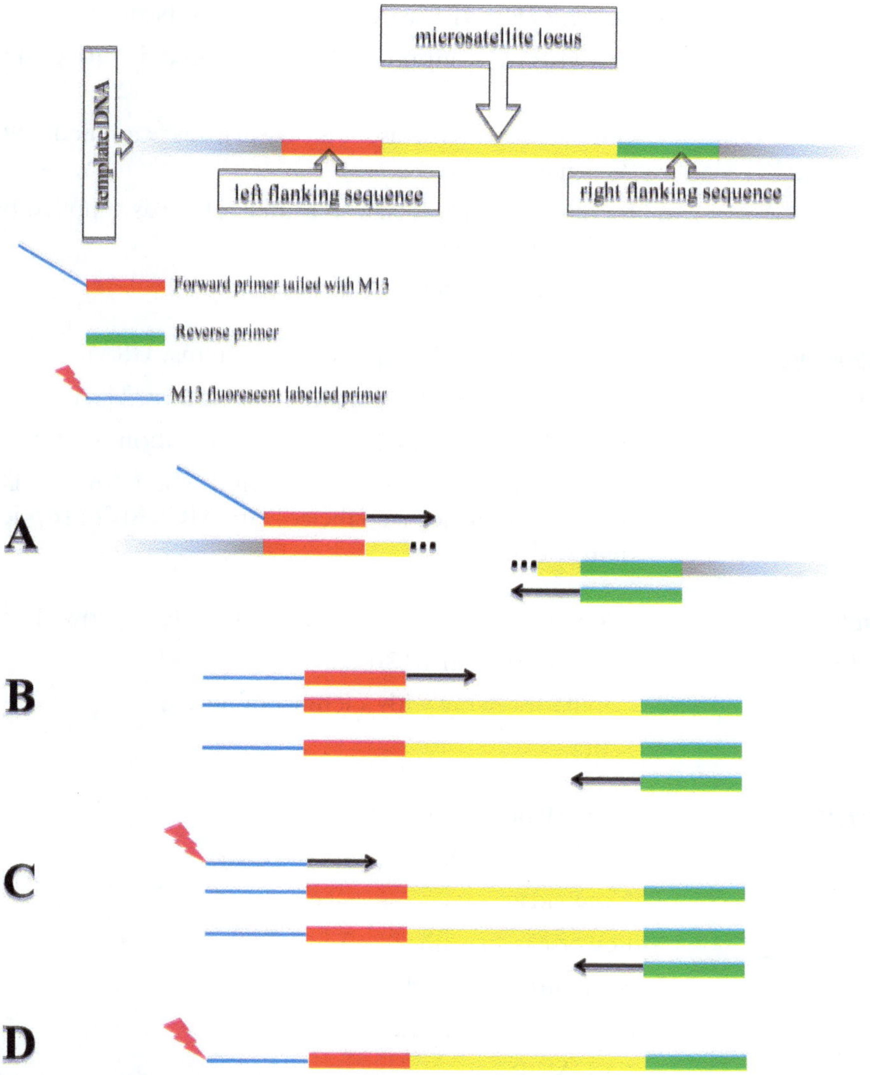

Fig. 1 The scheme of a single-tube, nested PCR performed with three primers. (**a**) The first cycle of the PCR reaction: the M13 tail is incorporated into the PCR product by the annealing and elongation of the forward primer. (**b**) During the subsequent cycles PCR products tailed with M13 are replicated. (**c**) The PCR products are the target for the fluorescently labeled M13 primer. During subsequent cycles fluorescently labeled M13 is used as the forward primer. (**d**) The final fluorescently labeled product can be analyzed on a laser detection system

2.4 PCR Components

1. Template DNA (50 ng/μl): stored in 100 μl aliquots at −20 °C.
2. PCR primers:
   (a) M13 primers (10 μM): stored in 100 μl aliquots at −20 °C.
   (b) Forward M13-tailed primers (2 μM): stored in 100 μl aliquots at −20 °C.
   (c) Reverse primers (10 μM): stored in 100 μl aliquots at −20 °C.

3. $MgCl_2$ (25 mM): stored in 1 ml aliquots at −20 °C.
4. dNTPs mix solution (10 mM): stored in 50 µl aliquots at −20 °C.
5. DSF *Taq* polymerase (5 U/µl): as supplied by Bioron (Germany), stored at −20 °C.
6. Incomplete NH_4 Reaction buffer (10×): as supplied by Bioron (Germany), stored at −20 °C.
7. "Ultrapure" water.

2.5 Electrophoresis Components

1. Hi-Di Formamide (Applied Biosystems, USA).
2. Running Buffer 10× (Applied Biosystems, USA).
3. NanoPOP-7 (MCLAB, USA) — separation matrix.
4. Orange DNA Size Standard (MCLAB, USA) — DNA Size Standard fluorescence-labeled with MCLAB's proprietary fifth orange dye.

2.6 Instruments/ Consumables

1. Greenhouse or growth chamber suitable to grow *Avena sativa*.
2. Genetic analyzer 3130XL.
3. Capillaries Array with a length of 36 cm.
4. Septa.

2.7 Software

1. GeneMapper® Software.
2. MS Excel 2010.
3. Notepad++.
4. Power Marker 3.5.
5. Structure ver. 2.3.4.
6. Structure Harvester.
7. CLUMPP 1.1.
8. DISTRUCT 1.1.
9. Arlequin 3.5.
10. GenAlex 6.5.

3 Methods

3.1 Growing the Plant Material

1. Conduct the germination in small pots with soil substrate for seeding. The germination can be conducted either in a greenhouse or in a growth chamber at ca. 20 °C and at least a 12 h photoperiod. The number of grains per pot should not exceed 15 per pot (Ø=9 cm) (*see* **Note 3**).

3.2 Collecting, Processing, and DNA Isolation from the Plant Material

1. Approximately 3–4 weeks after germination collect DNA samples separately from randomly chosen individuals of each accession number (*see* **Note 3**). Place about 100 mg of young, healthy leaf tissue of each plant in 2.0 ml tubes placed on ice.
2. Dry collected samples with a vacuum centrifuge or vacuum desiccator chamber filled with silica gel (*see* **Note 4**).
3. Ground the dried material in a bead mill homogenizer using stainless or glass beads 3 mm in diameter until a homogeneous powder is formed (*see* **Notes 5** and **6**).
4. Extract the total genomic DNA with a commercial kit(s) (*see* **Note 7**).
5. Determine DNA purity and concentration spectrophotometrically.
6. Determine DNA integrity on a 1.0% agarose gel (*see* **Note 8**).

3.3 Performing PCR

1. Thaw all PCR components at RT, mix them thoroughly (except for *Taq* polymerase), and pulse centrifuge briefly to collect the contents.
2. Carry out the reactions in a final volume of 10 μl.
3. Use the following final concentrations of the components and fill with water to the specified volume: 1× Taq buffer, 0.2 mM dNTP solution, 0.25 μM M13 primer, 0.25 μM reverse primer, 0.05 μM forward M13-tailed primer, 2.5 mM $MgCl_2$, 50 ng genomic DNA, 0.5 U DSF-Taq DNA Polymerase. Add all of the PCR components keeping the sample tubes in a PCR cooler.
4. Perform the amplification using the following temperature profile (Table 1).

3.4 Separation and Visualization

1. Prepare a mixture of Hi-Di Formamide and Orange DNA Size Standard in a 2.0 ml tube by adding 0.1 μl of the size standard and 10 μl of the formamide per each PCR sample. Subsequently, vortex thoroughly and pulse centrifuge briefly to collect the contents. Dispense the mixture into a 96-well plate, 10 μl per well.
2. Add 2 μl of post-PCR mixture to each well, mix thoroughly, and pulse centrifuge briefly to collect the contents. Heat the plate at 95 °C for 3 min to denature the DNA. Cool the samples for 2 min on ice. Seal the plate with Septa Seal (*see* **Note 9**).
3. Place the sample plate into the plate base and then snap the plate retainer onto them. Place it on the ABI 3130XL's autosampler.
4. Make sure there are no air bubbles around the capillary array knob or in the interconnect tube (*see* **Note 10**).
5. Set the instrument running conditions for the 36 cm capillary array, NanoPOP-7 polymer, and the following program:

Table 1
Temperature profile for the M13 tagged PCR reaction

Temperature (°C)	Time	Number of cycles	Comments
94	5 min		
94	30 s	15	
65	30 s		Touchdown 1 °C per cycle (65–50 °C)
72	45 s		
94	15 s	30	
50	15 s		
72	45 s		
72	10 min		
4	∞		

(a) Oven temperature: 60 °C.
(b) Polly fill vol: 6500 Steps.
(c) Current stability: 5.0 μAmps.
(d) PreRun voltage: 15.0 kV.
(e) Prerun time: 180 s.
(f) Injection voltage: 1.2 kV.
(g) Injection time: 23 s.
(h) Voltage number of steps: 20 *nk*.
(i) Voltage step interval: 15 s.
(j) Data delay time: 60 s.
(k) Run voltage: 15 kV.
(l) Run time: 1500 s.

6. Create and complete the plate record for autoanalysis.

3.5 Data Analysis

1. Score the results with the GeneMapper® Software according to the software manual.
2. Transfer the data into MS Excel.
3. Prepare the input file for Power Marker 3.5 [11].
4. Run Power Marker 3.5 and calculate the genetic diversity parameters such as:
   (a) Number of alleles per locus.
   (b) Allele frequency (*see* **Note 11**).

(c) Heterozygosity (*see* **Note 12**).

(d) Polymorphic information index (PIC) (*see* **Note 13**).

5. Prepare the input file for Structure ver 2.3.4 software [12] to assess the genetic structure by a model based approach (*see* **Note 14**).
6. Run Structure 2.3.4 software with the following parameters set (*see* **Note 15**):

   (a) The possibility of admixture.

   (b) Allele frequency correlated.

   (c) Run length 150,000 burning period length followed by 150,000 Markov Chain Monte Carlo (MCMC) replication.

   (d) Run each *k* value for ten times with the *k* value varying from 1 to 10.
7. Prepare a .zip file of the obtained results.
8. Open Structure Harvester [13] (http://taylor0.biology.ucla.edu/structureHarvester/#).
9. Upload the results .zip file.
10. A clear peak at the optimal *k* value will be shown.
11. Download the appropriate Clumpp input file.
12. Run CLUMPP 1.1 software [14] to find the optimal alignments of independent runs.
13. Use the output file directly into DISTRUCT 1.1 [15] for cluster visualization.
14. Prepare a bar plot (Fig. 2).
15. Prepare an input file for Arlequin 3.5 [16].
16. Run Arlequin 3.5 and perform:

   (a) Analysis of molecular variance (AMOVA) (*see* **Note 16**).

   (b) *F* statistics (*see* **Note 17**).
17. Prepare an input file for GenAlex 6.5 [17].

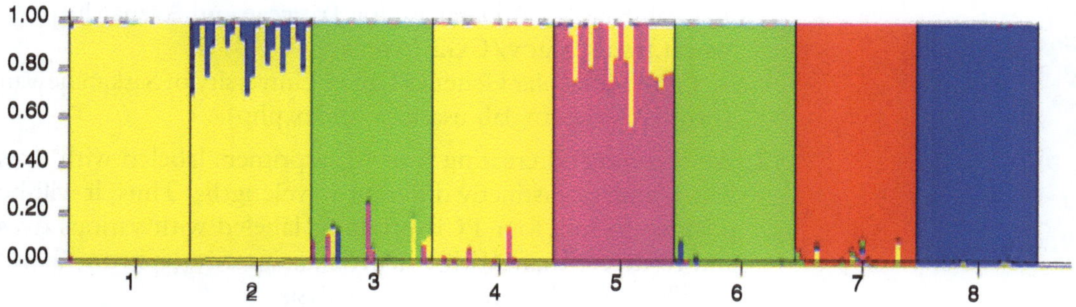

Fig. 2 A sample of the bar plot. Each *vertical thin bar* represents an individual within an accession that is marked by a *black border* and *number*

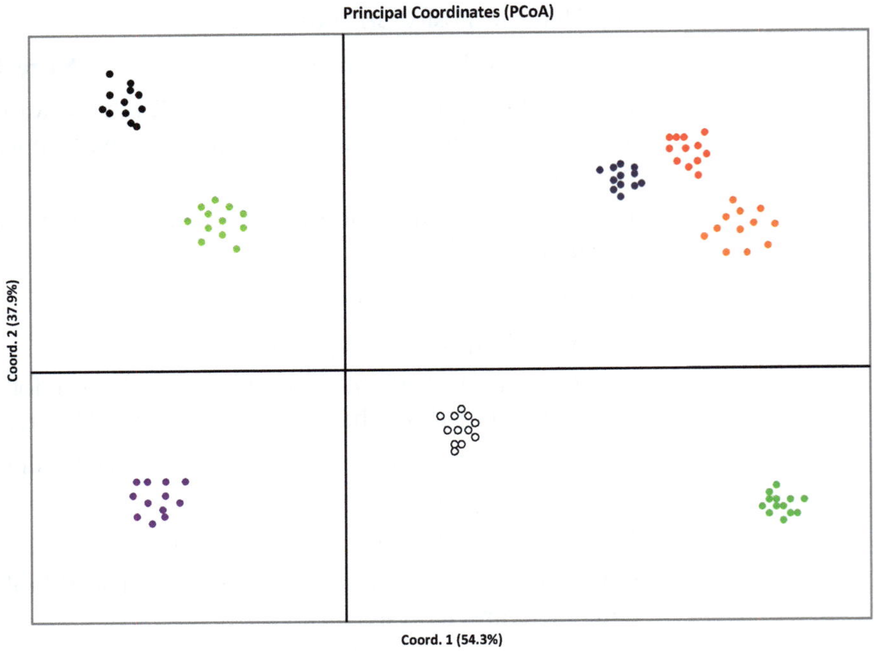

Fig. 3 A sample of the dot-plot of the first two coordinates. Each dot represents an individual. Each accession is marked by different color

18. Run GenAlEx 6.5 (*see* **Note 18**) and calculate the Nei [18] distance matrix.
19. Based on the distance matrix perform Principal Coordinate analysis.
20. Prepare a dot-plot of the first two coordinates to illustrate relation within and among accessions (Fig. 3).

4 Notes

1. The available *Avena* SSR markers database:

   GrainGenes – A Database for Triticeae and Avena (http://wheat.pw.usda.gov/GG2/Avena/)

   Crop Molecular Genetics Lab of University of Saskatchewan (http://www.cmglab.usask.ca/index.php).
2. We recommend creating four M13 primers labeled with dyes which emit at distinctly different wavelengths. Thus, it will be possible to mix four PCR products labeled with various dyes and to detect them with a genetic analyzer in one run. It will significantly reduce the cost of analysis.
3. The number of grains per accession depends on the experiment scale and internal diversity level.

4. After this step the dried material can be stored in a vacuum desiccator chamber filled with silica gel at ambient temperature for up to 1 month, or in tightly closed tubes at −80 °C for longer term storage. This step may be skipped if DNA is isolated directly from fresh tissue.
5. Fresh tissue may be homogenised with a bead mill only in the presence of an appropriate extraction buffer.
6. Stainless beads can be replaced by cheaper glass beads, but the number of beads must be increased. Both types of beads can be reused after treatment with a 10% sodium hypochlorite solution for at least 1 h and autoclaving.
7. We use the Genomic Mini AX Plant Kit from A&A Biotechnology but other commercial kits can be used.
8. If genomic DNA is smeared it does not necessarily mean total degradation. SSRs are shorter than 500 bp, so as long as the smear includes fragments larger than the predicted size of your amplicon, all will be well.
9. Septa Seal prevents damage to the capillary array; verify that the septa fits tightly onto the plate. Check the correct positioning of each sample – at the bottom of its well. No bubbles allowed.
10. In the case of air bubbles follow the bubble removal wizard.
11. Allele frequency represents the frequency of a particular allele for each marker.
12. Heterozygosity is the proportion of heterozygous individuals in the population.
13. Polymorphic information content that represents the amount of polymorphism within a population [19].
14. The parameter set can vary depending on the tested material.
15. Use Notepad++ to create an input file for Structure and any other text files.
16. The presence of molecular variance within and between the hierarchical population structure is assessed via the Analysis of molecular variance (AMOVA).
17. *F* statistics include FIT – deviations from the Hardy-Weinberg expectation across the whole population, FIS – deviation from the Hardy-Weinberg expectation within a population, and FST – correlation of alleles between subpopulations.
18. The GenAlEx is a MS Excel macro.

References

1. Soranzo N, Provan J, Powell W (1999) An example of microsatellite length variation in the mitochondrial genome of conifers. Genome 42(1):158–161
2. Powell W, Morgante M, McDevitt R, Vendramin G, Rafalski J (1995) Polymorphic simple sequence repeat regions in chloroplast genomes: applications to the population

genetics of pines. Proc Natl Acad Sci U S A 92(17):7759–7763
3. Tautz D (1989) Hypervariabflity of simple sequences as a general source for polymorphic DNA markers. Nucleic Acids Res 17(16):6463–6471
4. Gupta P, Balyan H, Sharma P, Ramesh B (1996) Microsatellites in plants: a new class of molecular markers. Curr Sci 45(1):45–54
5. Wang Z, Weber JL, Zhong G, Tanksley S (1994) Survey of plant short tandem DNA repeats. Theor Appl Genet 88(1):1–6
6. Li YC, Korol AB, Fahima T, Beiles A, Nevo E (2002) Microsatellites: genomic distribution, putative functions and mutational mechanisms: a review. Mol Ecol 11(12):2453–2465
7. Tautz D, Renz M (1984) Simple sequences are ubiquitous repetitive components of eukaryotic genomes. Nucleic Acids Res 12(10):4127–4138
8. Tautz D (1993) Notes on the definition and nomenclature of tandemly repetitive DNA sequences. In: Pena SDJ et al. (eds) DNA fingerprinting: state of the science. Springer, New York, pp 21–28
9. Eisen J (1999) Mechanistic basis for microsatellite instability. Microsatellites Evol Appl 1:34–48
10. Schuelke M (2000) An economic method for the fluorescent labeling of PCR fragments. Nat Biotechnol 18(2):233–234
11. Liu K, Muse SV (2005) PowerMarker: an integrated analysis environment for genetic marker analysis. Bioinformatics 21(9):2128–2129
12. Hubisz MJ, Falush D, Stephens M, Pritchard JK (2009) Inferring weak population structure with the assistance of sample group information. Mol Ecol Resour 9(5):1322–1332
13. Earl DA (2012) STRUCTURE HARVESTER: a website and program for visualizing STRUCTURE output and implementing the Evanno method. Conserv Genet Resour 4(2):359–361
14. Jakobsson M, Rosenberg NA (2007) CLUMPP: a cluster matching and permutation program for dealing with label switching and multimodality in analysis of population structure. Bioinformatics 23(14):1801–1806
15. Rosenberg NA (2004) DISTRUCT: a program for the graphical display of population structure. Mol Ecol Notes 4(1):137–138
16. Excoffier L, Laval G, Schneider S (2005) Arlequin (version 3.0): an integrated software package for population genetics data analysis. Evol Bioinform Online 1:47–50
17. Peakall R, Smouse PE (2012) GenAlEx 6.5: genetic analysis in Excel. Population genetic software for teaching and research-an update. Bioinformatics 28(19):2537–2539. doi:10.1093/bioinformatics/bts460
18. Nei M (1973) Analysis of gene diversity in subdivided populations. Proc Natl Acad Sci 70(12):3321–3323
19. Botstein D, White RL, Skolnick M, Davis RW (1980) Construction of a genetic linkage map in man using restriction fragment length polymorphisms. Am J Hum Genet 32(3):314

Chapter 13

Genotyping-by-Sequencing and Its Application to Oat Genomic Research

Yong-Bi Fu and Mo-Hua Yang

Abstract

Genotyping-by-sequencing (GBS) has emerged as a useful genomic approach for sampling genome-wide genetic variation, performing genome-wide association mapping, and conducting genomic selection. It is a combined one-step process of SNP marker discovery and genotyping through genome reduction with restriction enzymes and SNP calling with or without a sequenced genome. This approach has the advantage of being rapid, high throughput, cost effective, and applicable to organisms without sequenced genomes. It has been increasingly applied to generate SNP genotype data for plant genetic and genomic studies. To facilitate a wider GBS application, particularly in oat genetic and genomic research, we describe the GBS approach, review the current applications of GBS in plant species, and highlight some applications of GBS to oat research. We also discuss issues in various applications of GBS and provide some perspectives in GBS research. Recent developments of bioinformatics pipelines in high-quality SNP discovery for polyploid crops will enhance the application of GBS to oat genetic and genomic research.

Key words Next-generation sequencing, Genotyping-by-sequencing, Single nucleotide polymorphism, Contig, Oat, Genetic diversity, Linkage mapping, Associated mapping, Genomic selection

1 Introduction

Molecular markers, revealing polymorphisms at the DNA level, are important genetic tools in genetics and genomics [1, 2]. Many genetic researches in genetic diversity, association mapping, and genomic selection are highly dependent on effective genome sampling with sufficient and informative genetic markers such as single nucleotide polymorphism (SNP) [1, 3]. However, many species of interest are lacking of SNP markers and related genetic investigations are hindered. Previous efforts have developed SNP markers through various approaches such as chip hybridization or targeting specific genomic regions [4, 5]. However, such efforts for SNP discovery are expensive and labor intensive, as plants usually have large complex genomes with abundant sequence repeats and genome duplications [6]. Also, many species do not have sequenced

Sebastian Gasparis (ed.), *Oat: Methods and Protocols*, Methods in Molecular Biology, vol. 1536,
DOI 10.1007/978-1-4939-6682-0_13,

genomes and are considered to be nonmodel plants, making the SNP discovery more challenging [7].

Genotyping-by-sequencing (GBS) has emerged as a promising genomic approach for exploring plant genetic variation on a genome-wide scale [8–12], thanks to the advances in next generation sequencing (NGS) technologies [13, 14]. The GBS approach is based on genome reduction with restriction enzymes (REs) [15]; does not rely on a reference genome for SNP discovery; is a combined one-step process of marker discovery and genotyping; and provides a rapid, high throughput, and cost effective tool for a genome-wide analysis of genetic variation for a range of nonmodel species and germplasm sets [16, 17]. These characteristics are advantageous and encouraging for many genetic analyses of plants with no or insufficient SNP markers available [18].

The GBS approach has been successfully applied for genetic mapping [9, 19, 20], genomic selection for breeding [21, 22], and genetic-diversity studies [10, 17, 23]. However, GBS is still evolving to address key technical issues on uneven genome coverage and large missing SNP data, and many uncertainties and challenges remain in its applications [24–27], particularly for polyploid plants [28, 29]. Also, most GBS protocols were developed largely based on model plants (those with sequenced genomes), and little is known about its applicability or feasibility for nonmodel plants [7, 18, 30]. Existing bioinformatics pipelines need to be modified for generating de novo contigs to be used as a reference for SNP discovery and genotyping [31]. Moreover, GBS is conceptually simple, but in practice involves many steps requiring a range of molecular biology skills and different bioinformatics analyses [30, 32]. All of these issues may have hindered its applications, particularly for researchers with little experience in NGS research and bioinformatics analysis.

This chapter is written with the hope to facilitate GBS applications and explore its full potential to improve genomic research in oat and other crops. Specifically, we will describe the GBS approach, review the current applications of GBS in plant species, and highlight some applications of GBS to oat research. We also discuss issues in various GBS applications and provide some perspectives in GBS research.

2 Genotype-by-Sequencing

2.1 Background

It has been technically difficult to develop sufficient SNP markers for a plant species [1, 4], until recent advances in NGS technologies [14, 16], making it possible to sequence the whole genomes of some plant species [13]. Among the whole genome resequencing efforts is the application of NGS to rice (*Oryza sativa*; [8]) and maize (*Zea mays*; [33]). Inspired by these whole genome resequencing efforts and built upon the protocol of restriction site associated DNA (RAD) tags [19], Rob Elshire and his colleagues at Cornell University developed and first coined the GBS approach as a simple highly multiplexed system for constructing reduced representation libraries for Illumina sequencing [9]. Elshire's system was also equipped with a bioinformatics pipeline for SNP discovery and genotyping. This system was first demonstrated in maize and barley (*Hordeum vulgare*) and generated thousands of SNP markers for use in genetic analyses. However, Elshire's GBS protocol is not unique, and many similar GBS protocols were also developed, following the same idea of genomic reduction [15], by taking advantage of NGS technology [20, 34, 35]. For example, we also developed a similar GBS protocol using the Roche 454 pyrosequencing platform [10, 36] that was capable of generating thousands of genotyped SNP markers specifically for plant genetic diversity analysis [17]. Efforts to expand GBS protocols and evaluate them in different organisms for various genetic analyses such as genome-wide association mapping and genomic selection have also been fruitful [21, 37, 38]. These developments have shown the promise of the GBS approach in SNP discovery and genotyping [11].

2.2 Concept, Procedure, and Illustration

Conceptually, GBS is simple and can be regarded as one of several reduced-representation sequencing methods for genotyping [32]. The reduced-representation or genome reduction methods are based on the digestion of the genome by REs, followed by selection of a subset of digested DNA fragments to sample various regions of the genome. The selected DNA fragments are ligated with adaptors and different samples are labeled with barcodes. The barcoded DNA fragments (or libraries) are sequenced using a NGS platform capable of sequencing a huge number of fragments each with multiple sequence reads [9, 19]. The acquired sequence reads are analyzed for SNP discovery and genotyping by mapping reads to the reference genome of the species using specific bioinformatics pipelines or software [39–41]. For a species without a reference genome, consensus contigs are generated de novo from raw sequences and used as a pseudo-reference genome [6, 32]. Thus, GBS could be regarded as a low-genome-coverage genomic tool for sampling genome-wide genetic variants.

Generally, a GBS application is still a complex process and can involve several major components as sample preparation, library assembly, sequencing, SNP calling, and genetic analysis. Major steps include DNA extraction and digestion, adapter ligation, PCR amplification, fragment size selection, library pooling, sequencing, data processing, SNP calling, and genetic or genomic analysis [32]. These steps may vary among different GBS protocols in RE use, NGS platform, and in bioinformatics analysis for SNP genotyping for different study objectives. For example, some GBS protocols employ two REs for DNA digestion (e.g., [34]), while some use single RE (e.g., [9]). Each published protocol has its own detailed GBS steps, and a researcher needs to follow those steps precisely to generate GBS sequence data. Based on our experience, it is not difficult to apply a given GBS protocol to generate required GBS DNA libraries for sequencing on a NGS platform, provided that appropriate validation is done throughout the protocol.

To illustrate a complete GBS procedure, we describe the diversity-focused GBS (gd-GBS) protocol we had developed for plant genetic diversity [32]. This protocol has five components and each component has many major steps, as illustrated in Fig. 1. For example, the Sample Preparation component involves seven major steps for DNA extraction, quantification, restriction digest, ligation, sample clean-up, PCR amplification, and quantification. Similarly, the SNP Calling component has five major steps for collapsing reads, assemble contigs, align reads to contigs, summarize alignments and call SNPs, and generate and format genotype data. The details for each step are described in the easy-follow, step-by-step lab guide. The major features of gd-GBS are the use of two restriction enzymes and fragment size selection to reduce genome complexity, the application of Illumina multiplexing indexes for barcoding and the availability of a custom bioinformatics pipeline for genotyping a diploid species. The complete gd-GBS protocol, including the bioinformatics pipeline "npGeno" (short for nonmodel

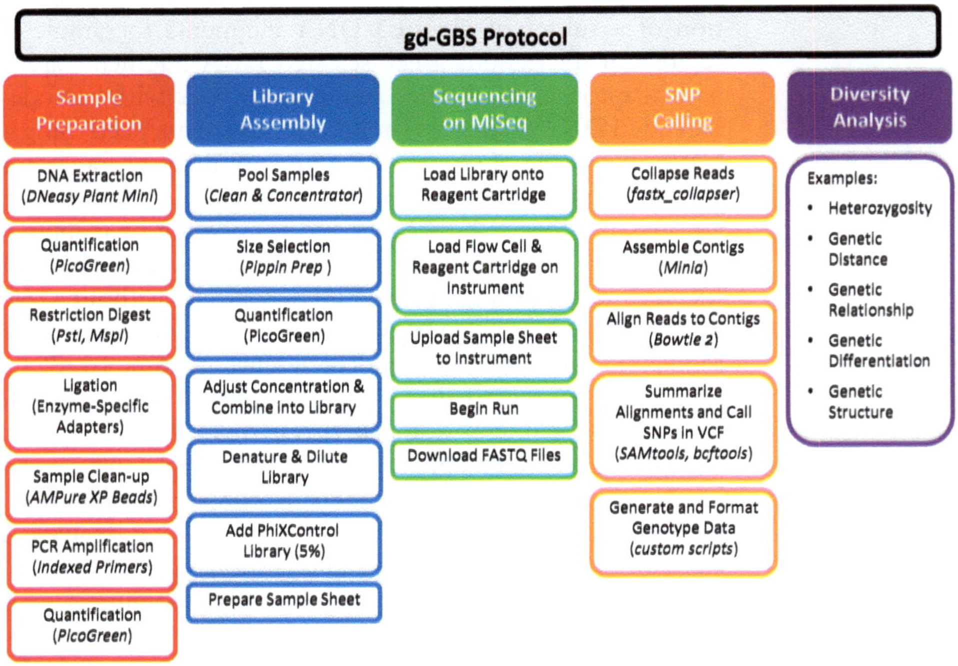

Fig. 1 Flow chart of the genetic diversity-focused genotyping-by-sequencing (gd-GBS) protocol. It consists of five major parts and many steps. Arranged in *columns* beneath each part are the ordered major steps in *outlined boxes* of the same color. *Bracketed terms* refer to specific equipment, reagents, kits, or software used (adapted from Peterson et al. [32])

plant genotyping), is fully accessible for use [32]. The use of the gd-GBS protocol, however, assumes a researcher has general knowledge of plant genetic diversity analysis, can perform a DNA library preparation, has access to sufficient computing resources (e.g., a Linux server), and has the basic operational skills to work in a Unix-like environment.

2.3 Expanded GBS Protocols

Since the publication of the genome reduction method [15], efforts have been made to sequence multiplex DNA libraries through the combination of NGS with DNA barcoded adapters. One of the significant developments was the publication of complexity reduction of polymorphic sequences (CRoPS) [42] and restriction association DNA tag (RAD-seq) [19] protocols. Compared to RAD-seq, the original GBS library preparation is greatly simplified with less DNA, does not require random shearing and size selection, and is completed in only two steps in PCR plates, followed by PCR amplification of the pooled library [9]. Since 2011, many GBS protocols have been developed with the focus on optimizing library and fragment sizes for sequencing. Table 1 shows some representative GBS protocols published up to 2015. Among those, they are dd-RAD [34], 2b-RAD [35], ezRAD [43], RESTseq [44], gd-GBS [32], and MIG-seq [45].

These protocols have the similar features in the use of frequent-cutting enzymes to generate fragments of appropriate length and sample-specific barcodes for multiplexing sequencing. However, variation also exists in the selection of RE and the fragments size range and in the use of platform-specific barcodes. For example, RAD-seq [19] uses one enzyme for DNA digestion and applies a sonication step to shear DNA after digestion for fragment size selection while the original GBS [9] relies on a PCR step to preferentially amplify shorter fragments over longer fragments. 2b-RAD was designed with type-IIB of RE that cuts DNA into small, uniformly sized fragments (33–36 bp) suitable for short-read sequencing. gd-GBS adopted two enzyme systems from dd-GBS to focus on wider genome sampling for a genetic diversity analysis [32]. More recently, MIG-seq was developed as a primer-based, rather than RE-based, size selection in genome reduction to sample genomic variants [45]. For many GBS protocols listed in Table 1, the number of unique fragments can be optimized by using different REs with desirable cut frequencies in a genome. dd-RAD and ezRAD are unique in selecting DNA fragments for sequencing through the size selection step in the library preparation. For this, GBSX was developed to assist the RE selection and the barcoding design [46]. Recently, we explored better RE combinations through an in silico analysis and empirical validation and found a new set of 21 RE combinations that can be applied to enhance GBS applications with increased genome sampling and improved SNP genotyping [47]. These GBS expansions make the GBS approach more versatile for wide applications in genomic studies.

Table 1
Representative GBS and related protocols published in peer-reviewed journals

Protocol	Enzyme	Fragment size	Sequencing platform	Reference
CRoPS (Complexity Reduction of Polymorphic Sequences)	*Hpa*II and *Mse*I	250–500 bp	Roche 454	[42]
RAD-seq (Restriction site Associated DNA sequencing)	*Sbf*I or *EcoR*I	Size-selection	Illumina	[19]
GBS (Genotyping-by-sequencing)	*ApeK*I	~350 bp	Illumina	[9]
Double-digested GBS	*Pst*I and *Msp*I	~350 bp	Illumina	[20]
Double-digested RAD-seq	*EcoR*I and *Msp*I	Size-selection	Illumina	[34]
SBG (Sequence-based genotyping)	*EcoR*I and *Mse*I	Size-selection	Illumina	[94]
Restriction enzyme sequence comparative analysis	*Mse*I and *Nla*III	Size-selection	Illumina	[95]
2b-RAD	*BsaX*I or *Alf*I	33–36 bp	SOLiD/Illumina	[35]
ezRAD	*Mbo*I and *Sau3A*I	400–500 bp	Illumina	[43]
REST-seq (Restriction fragment sequencing)	*Taq*I and *Tru*I	70–105 bp	Ion Torrent	[44]
gd-GBS (genetic diversity-focused GBS)	*Pst*I and *Msp*I	~350 bp	Illumina Miseq	[32]
ddRAD	*Pst*I and *BamH*I	Size-selection	Ion Torrent	[96]
MIG-seq (multiplexed ISSR genotyping by sequencing)	*PCR* Primer-*based*	Size-selection	Illumina MiSeq	[45]

2.4 Bioinformatics Tools

Acquiring GBS sequence data has become more feasible than before with the advance of NGS platforms such as Illumina HiSeq [14] and more cheaper and efficient with the third generation sequencing platforms like PacBio SMRT and Oxford Nanopore [48]. Thus, the need for effective bioinformatics tools has become more critical in GBS applications [49]. The last several years have seen considerable efforts in the development of bioinformatics tools for analyzing GBS sequence data. Table 2 lists representative bioinformatics tools publicly available for GBS applications, including STACKS [40], UNEAK [23], TASSEL-GBS [41], npGeno [32], PyRAD [50], SWEEP [51], RIG [52], Haplotag [53], and GBSTools [54]. Most pipelines can identify loci, either de novo or reference based, and call SNP genotypes in a set of assayed individuals. These tools are publicly accessible and capable of analyzing different GBS sequence data for different research objectives.

Table 2
Representative bioinformatic pipelines publicly available for GBS data analysis

Pipeline	Open resources	SNP call on	Ploidy	Reference
STACKS	http://www.creskolab.uoregon.edu/stacks	Reference or de novo	Diploid	[40]
UNEAK	http://www.maizegenetics.net/gbs-bioinformatics	de novo	Polyploid	[23]
TASSEL-GBS	http://www.maizegenetics.net/tassel	Reference or de novo	Diploid	[41]
IGST-GBS		Reference	Diploid	[97]
RADtyping	http://www2.ouc.edu.cn/mollusk/detailen.asp?id=727	de novo	Diploid	[98]
PyRAD	http://www.dereneaton.com/software	de novo	Diploid	[50]
dDocent	http://dDocent.wordpress.com	de novo	Diploid	[99]
npGeno	http://www.mdpi.com/1424-2818/4/6/665/s1	de novo	Diploid	[32]
SWEEP	https://github.com/jclev-uga/SWEEP	de novo	Polyploid	[29]
paSNPg		de novo	Diploid	[31]
AftrRAD	https://github.com/mikesovic/AftrRAD.git	de novo	Diploid	[100]
RIG	https://github.com/MulletLab/RIG	Reference	Diploid	[52]
Haplotag	http://haplotag.aowc.ca	de novo	Polyploid	[53]
GBSTools	http://med.stanford.edu/bustamantelab/software.html		Diploid	[54]

Among different pipelines, STACKS and UNEAK have played an important role in GBS data analysis. UNEAK was designed to operate independently from a reference genome while STACKS may be used for SNP discovery with or without a reference genome. TASSEL-GBS was originally tailored to the GBS protocols of Elshire et al. [9] and Poland et al. [20] and was specifically designed to analyze large GBS data using a reference genome. However, npGeno was developed to analyze GBS sequence data for non-model plants from de novo assembled contigs. Most pipelines, including those mentioned above, are applicable only to diploid species, but UNEAK, SWEEP, and Haplotag were developed for polyploid genomes. Also, most pipelines require the installation of the third party programs (e.g., to assemble or align sequences) while UNEAK requires only the installation of a JAVA run-time environment [23], which makes UNEAK popular with researchers with access to restricted or limited computer platforms. TASSEL-GBS can operate on desktop or laptop computers. Moreover, some pipelines vary according to the library preparation method or sequencing platform used, and different pipelines use their own modular data mining strategies. GBStools was developed with a statistical method to remove genotyping errors caused by restriction site polymorphism for better genotyping accuracy [54].

The basic workflow in those pipelines for variant discovery from raw sequence data can be divided into three sequential steps: (1) raw data processing, including separating reads into their corresponding samples based on barcode matching, evaluating and selecting high quality reads, and cleaning up short reads, (2) read alignment to a reference genome or de novo assembly of sequence tags, and (3) variation discovery and annotation [49]. To illustrate these steps, the flow chart of npGeno is shown in Fig. 1 [32]. The script *npGeno.sh* consists of four shell scripts that automate freely available software and custom Perl scripts. The first constructs contigs from sequence reads from all samples; the second calls SNPs using the constructed contigs as a reference; the third filters resulting SNPs; and the fourth formats data outputs. To construct reference contigs, Fastx_collapser [55] is used to collapse all identical reads to separate unique sequence data, and Minia software [56] is used to construct de novo contigs. Bowtie 2 [57] is employed to map the reads from each sample against the reference contigs. SAMtools [39] is used to create a pile-up file summary of the aligned reads relative to the contigs, and BCFtools calls SNPs in a variant call format (VCF) file. Custom Perl scripts are also developed to create tab-delimited genotype and haplotype data from the VCF files, to remove duplicates and missing data, and to reformat output data required for various diversity analyses. The pipeline takes FASTQ input, along with three other input files, and outputs seven data files for different genetic analyses. It was developed for use on a Linux operating system.

The accuracy of SNP calls may vary among the listed bioinformatics tools and little comparison has been made among these published tools. Issues associated with the use of these tools also exist. Converting the raw GBS sequences into SNP genotype data involves a number of steps, each of which contributes to the accuracy of the final genotype calls [58]. Thus, separating high-quality markers from sequencing noise is challenging [16]. Methods appropriate for one sequencing platform are not necessarily appropriate for those from other NGS platforms, and new library preparation methods may produce novel sources of noise for SNP calls. Many sensitive and specific statistical models are required for marker discovery and genotyping using NGS data [58, 59], particularly in polyploidy species. The analysis of NGS data is also a fast-evolving field, and improving genotyping accuracy is an active research area for the GBS bioinformatics analysis.

3 Application of GBS in Major Crop Plants

To understand how widely the GBS approach has applied to sample genome-wide SNPs, we conducted a simple literature search. It was found that extensive GBS applications have emerged rapidly after the publication of the original GBS protocol [9]. The search with the key word of *Elshire GBS* in Google Scholar between 2011 and 2015 revealed 978 publications citing Elshire et al. [9]. Scanning those 978 citations, we found 531 papers had GBS applications in plants. Sorting these GBS applications based on publication year showed a dramatic increase in GBS applications from 2011 to 2015 (Fig. 2a). This preliminary literature survey clearly demonstrates the increasing role of the GBS approach in genetic or genomic studies of plant species.

Further examination on the application types revealed a variety of uses for GBS generated data, ranging from SNP discovery to linkage mapping, QTL identification, genome-wide association mapping (GWAS), genomic selection (GS), genetic diversity analysis, and phylogenetic analysis (Fig. 2b). Linkage map construction accounted for 134 papers between 2011 and 2015, followed by GWAS, QTL identification, and SNP discovery with 116, 69, and 67 papers, respectively. This result indicates the dominant role of the GBS approach in the development of SNP markers for linkage map construction. This is understandable as a linkage map is the fundamental tool for downstream genetic analyses such as QTL identification and gene cloning. Also, the GWAS analysis is greatly enhanced with the GBS applications for genome-wide SNP genotyping. For example, the GWAS analysis in *Arabidopsis thaliana* [60] had identified many true genotype-phenotype associations

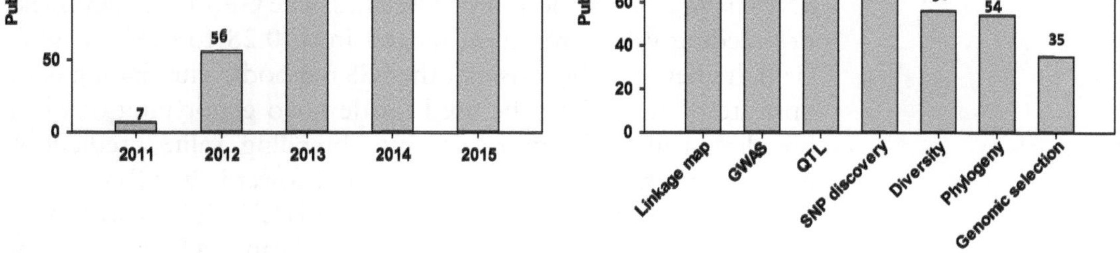

Fig. 2 Publication counts with GBS applications in plant species between 2011 and 2015 with respect to publication year (**a**) and application type (**b**)

and causal loci at a higher resolution than the classical QTL analysis through linkage mapping. Moreover, there were extensive GBS applications for SNP discovery, particularly in nonmodel species. These GBS applications helped to generate genome-wide genetic variants for genetic diversity analysis (e.g., [17]) and phylogenetic analysis of important plant species with complex genomes (e.g., [61]).

The application of GBS has been reported in several major crop species, including maize [62], rice [8, 63], wheat (*Triticum aestivum*; [20, 21]), and soybean (*Glycine max*; [64]). The applications have been focused on the SNP discovery, linkage mapping, QTL analysis, GWAS, and GS. Reports in maize, soybean, and wheat have demonstrated the effectiveness and feasibility of the GBS approach to infer the genotype-to-phenotype associations for many agronomic traits [21, 62, 64]. The highly dense genome-wide SNP markers acquired through GBS in these crop species have increased the resolution to characterize the genetic architecture of complex traits and allowed for proper evaluations on the effectiveness of the GBS applications in improving complex traits of agronomic importance [11, 18].

GWAS utilizes the genetic association mapping principle on the genome scale to identify statistically significant phenotype-genotype associations. Since the first GBS application for GWAS in lodgepole pine (*Pine contorta*) [37], many more GWAS reports have been found in crop species, including maize [65], rice [63], sorghum (*Sorghum bicolor*; [66]), perennial ryegrass (*Lolium perenne*; [67]), *Brassica napus* [68], *Miscanthus sinensis* [69], and Reed Canarygrass (*Phalaris arundinacea*; [70]). The use of GBS for GWAS makes it possible to genotype larger plant populations with a higher density of SNP markers than was previously possible. With larger populations, more recombination breakpoints are identified, allowing for the positioning of candidate genes with higher precision. The realization of low-cost genetic markers through GBS approach has paved the way for applying genomics-assisted breeding in crops breeding programs [18, 71].

The GBS approach has also played an increasing role in GS for crop breeding and improvement. Poland et al. [21] applied the GBS approach for GS in wheat and identified a new set of 41,371 SNPs from 254 advanced breeding lines. The GBS-based accuracies of breeding value prediction ranged from 0.28 to 0.45 for grain yield, higher than those using other GS methods. This finding demonstrates that GBS can be used for de novo genotyping of wheat breeding panels for more accurate breeding value predictions. More research efforts are being directed toward the GBS applications for GS in different breeding programs [71]. The marked GBS applications for GS were also found in soybean [64], pea [72], and tetraploid alfalfa [73].

4 Application of GBS to Oat Research

Cultivated hexaploid oat [*Avena sativa*; $2n=6x=42$; 1C-value = 13.2 (pg) DNA] is a nutritionally important cereal crop produced for both a source of food and feed [74]. To meet the agronomic and nutritional needs in oat, continuous genetic improvement is required, particularly through molecular breeding. Different oat DNA markers such as DArT and SNPs have been developed and widely explored in various oat genetic analyses such as linkage mapping, QTL identification, functional gene mapping, GWAS, and GS [75–81]. However, highly dense genome-wide SNP markers across the oat genome are largely lacking [82]. Oat genomic research had been lagged behind the other major crops such as rice and maize, due to the large size and complexity of its hexaploid genome, numerous chromosomal rearrangements, and insufficient sequence information [80]. The GBS approach has the potential to accelerate oat genetic and genomic research [83].

4.1 Marker Development

A large genome and high genome complexity makes whole genome sequencing in oat particularly challenging. An alternative path is to develop the highly dense consensus linkage map with huge number of SNP markers anchored on chromosomes. One of the significant efforts is the application of RNA-seq, along with an assembly pipeline that takes the complex oat genome into consideration and is designed for Roche 454 cDNA sequencing reads [79]. Based on cDNA sequencing in four cultivated oat genotypes, about 127,000 contigs were assembled. The largest number of contigs (42,147) in the genotype TAM was used as the reference assembly for in silico SNP detection. A total of 9448 candidate loci were obtained. This effort helped to demonstrate the effectiveness of RNA-seq approach to acquire cDNA-associated SNPs in a crop with complex genome. RNA-seq has the advantage of minimizing the impacts of the genome complexity with paralogous and orthologous gene duplication, chromosomal rearrangements, and polyploidy [84].

Oliver and her colleagues continued the RNA-seq research to construct the first complete, physically anchored consensus map of hexaploid oat [85]. Using an advanced SNP discovery method for complex genomes, they discovered about 11,000 highly confident in silico SNPs. GoldenGate genotyping with 3072 SNP assays yielded 1311 robust markers, of which 985 were mapped in 390 recombinant-inbred lines from six biparental mapping populations ranging in size from 49 to 97 progeny. The constructed consensus map included 985 SNPs and 68 previously published markers, resolving 21 linkage groups with a total map distance of 1838.8 cM. Through SNP deletion analysis of chromosome-deficient monosomic hybrid stocks, the consensus linkage groups were assigned to 21 chromosomes. Comparison on the alignments between oat and sequenced genomes of rice and *Brachypodium* displayed large regions of macro genome colinearity among these three species. The constructed physically anchored consensus map and the orthology-based genomic discovery established a fundamental base to accelerate oat genomic research.

Simultaneously, considerable efforts have also been made based on a variety of bioinformatics analyses of cDNA and genomic DNA sampled from 20 or more diverse oat cultivars to develop a 6K BeadChip SNP assay as a reliable set of reference SNPs for molecular breeding [82]. The 6K assay contains 257 Infinium I and 5486 Infinium II designs corresponding to 5743 SNPs. Of those, 4975 SNPs yielded successful assays after array manufacturing. These SNPs were validated in 1100 samples from six recombinant inbred line mapping populations and sets of diverse oat cultivars and breeding lines. It was found that about 3500 SNPs followed discernible Mendelian segregations. This 6K array, along with the saturated linkage map [85], has started to play a role in various oat genomic analyses such as QTL mapping and gene identification (e.g., [86]).

4.2 GBS Application

Huang et al. [83] reported the first application of GBS in oat to develop genome-wide SNP markers from genomic DNAs. They employed the GBS protocol of Poland et al. [20], constructed 95-plex libraries from seven biparental mapping populations and diverse inbred lines from breeding programs around the world, generated GBS sequence data using Illumina HiSeq 2000, and analyzed GBS data using the UNEAK pipeline. The application generated a huge set of 45,117 SNPs and the detected markers were integrated into the oat consensus map [85]. To saturate the consensus map with more markers and correct the marker mapping ambiguities, Chaffin et al. [87] made further GBS application to genotype progeny from 12 biparental recombinant inbred line populations. More than 7202 integrated markers spanning 2843 cM of the consensus map had drastically improved the resolution of the previous linkage map with more accuracy and better genome sampling. They also assessed various technical issues associated with the GBS application to hexaploid oat and demonstrated that GBS is a powerful and cost-effective approach for acquiring genome-wide genetic variants for oat breeding and genomic studies.

Further research efforts are being made through GBS on oat QTL analysis, GWAS, and GS (Nicholas Tinker, personal communication). Following the GBS application of Huang et al. [83], Klos et al. [88] explored the population structure, linkage disequilibrium (LD), and genotype-phenotype association with heading date in 635 oat lines based on 4561 integrated SNP loci. Association with heading date was found on linkage groups Mrg02, Mrg12, Mrg13, and Mrg24. However, the application of GBS to assess oat genetic diversity [89], explore exotic gene pools [90], and infer genome evolution [91] has not been reported yet, but should be a fruitful area of research.

One major challenge facing with the oat GBS application is the polyploidy of oat genome, complicating the SNP genotyping among homoeologous sub-genomes. With recent development in bioinformatics tools for SNP genotyping in polyploid plants, the accuracy of SNP genotyping should be increased in any oat GBS application. However, little attention has been paid to assess the effectiveness of these bioinformatics tools in oat SNP genotyping. To advance the haplotype inference in a polyploid plant, Tinker et al. [53] also developed and demonstrated a bioinformatics tool *Haplotag* to capture SNP haplotype in complex oat genome. These efforts, along with the other released pipelines such as SWEEP, will enhance the genomic research in oat and other polyploid plants.

5 General Discussion

Reviewing the application of GBS in major crop plants reveals an increased use of GBS in various crop genomic studies. It also demonstrates the effectiveness of the GBS approach in sampling genome-wide genetic variants. To facilitate the wider application of GBS, we provide some practical considerations on GBS applications and present our perspectives on further GBS research from our own GBS research and experience.

5.1 Practical Considerations

To make GBS applications more effective, several factors need to be considered in the configuration of GBS-assisted research: the GBS genotyping objective, lab bench procedure, sequencing platform and cost, bioinformatics analysis, and downstream genetic analysis. These factors have been thoroughly discussed (*see* refs. 25, 32, 92). The objective of a GBS genotyping is the most important factor, as different genetic analyses would require different experimental design and different levels of genomic coverage and/or genotyping error tolerance. GWAS and GS, for example, usually involve complicated study materials or mapping populations, but each requires different level of genomic coverage or genotyping error tolerance. In contrast, genetic diversity and population genetic analysis would include a large number of samples from nonmodel species dealing with de novo sequencing and SNP calling with the associated lower genome coverage. As the GBS protocols vary mainly in restriction digest and size selection and may include different bioinformatics pipelines for SNP genotyping, selecting a GBS protocol should take into consideration their advantages and limitations, particularly with respect to genome coverage and read depth. Clearly, the two-enzyme GBS protocol has become more effective than the others in sampling genome variants [47]. Many bioinformatics tools are developed, but little effort has been made to compare their effectiveness with respect to SNP count and accuracy, run time, memory use, operation system, and usage complexity. Thus, little guidance can be derived in the selection of bioinformatics tools specifically for a GBS data analysis.

Several practical issues are also worth discussion. First, DNA purity is critical for enzyme digestion and DNA library preparation for DNA sequencing. This issue has been rarely raised in any GBS applications. Second, bioinformatics analysis of huge GBS sequence data will become the dominant task in any GBS applications. Many user-friendly bioinformatics pipelines have been developed for those researchers with little experience in bioinformatics. Applications of these bioinformatics tools with default settings may not be optimized and would carry hidden genotyping errors. However, it is usually considered good practice to first run the data with default parameters to use as a baseline to compare the response of the algorithms to any setting changes, since a small setting change may result in substantially different output data [93]. Third, the application of GBS has generated large amount of SNP data, but few applications have empirically verified the accuracy of SNP genotyping. Fourth, the large amount of GBS SNP data acquired from a project has become an issue for downstream genetic analyses, as traditional genetic analysis software and/or the computing platform originally designed for lacks the capacity to handle the data volume. For example, in population genetics, GenAlEx can only handle up to 8000 SNPs and STRUCTURE would require weeks or months to analysis more than 10,000 SNPs. Thus, related software for various genetic analyses needs to be expanded. Fifth, computational resources also need to be expanded to keep pace with the accumulating huge GBS sequencing datasets, as desktop computers will be limited in the number of samples they can process by the available RAM memory [49]. Thus, the access to a computing facility, such as Linux servers, and some operational skill in a UNIX-like environment, is required for a researcher professionally analyzing the large GBS data.

5.2 Future Directions in GBS Research

The described GBS approach is not lacking of technical limitations and issues, and will continue to evolve. Our review revealed that many key technical issues remain to be resolved for more effective application of GBS in plant genomics studies and plant breeding practices [18, 71]. The issues are associated with uneven genome sampling, PCR artifacts, low genotyping accuracy, and large missing data [47]. An informative GBS application requires genome-wide SNP discovery with desired genome coverage, high genotyping accuracy, more balanced observation and less bias introduced from sequence mapping, heterozygote detection, and data filtering.

Efforts are still needed to modify existing or develop new robust GBS protocols for desired genome coverage, new bioinformatics pipelines for SNP discovery and genotyping, particularly in polyploid plants, and effective imputation methods for large missing data. Some focus could also be placed on the effectiveness of sampling non-repetitive genomic regions, the acquisition of fewer SNP markers with higher sequence coverage for better SNP accuracy, the increased multiplexing for large population studies, and new genomic applications. The genetic diversity analysis generally requires wider genome coverage with more markers to sample genetic background, while association mapping and genomic selection require higher marker accuracy to acquire genic effects across the genome. Both genome coverage and genotyping accuracy are essential for downstream genetic or genomic analyses. Thus, different GBS protocols should be developed for different research purposes.

6 Conclusive Remarks

Genotyping-by-sequencing has emerged as a useful genomic tool for exploring genome-wide genetic variation. This approach has the advantage of being rapid, high throughput, cost effective, and applicable to organisms without sequenced genomes. It has been increasingly applied to generate SNP genotype data for plant genetic diversity analysis, linkage mapping, genome-wide association mapping, and genomic selection. Recent developments of bioinformatics pipelines for high-quality SNP discovery in polyploid plants will enhance the application of GBS to oat genetic and genomic research.

Acknowledgment

We would like to thank Gregory Peterson, Yifang Tan, Sateesh Kagale, and Nicholas Tinker for their helpful comments on the early version of the manuscript. This work was supported by an A-Base research project of Agriculture and Agri-Food Canada to Y.B.F. and National Natural Science Foundation of China (Grant No. 31200481) and the China Scholarship Council Postdoctoral Abroad Grant to M.-H.Y.

References

1. Rafalski A (2002) Applications of single nucleotide polymorphisms in crop genetics. Curr Opin Plant Biol 5:94–100
2. Nybom H, Weising K, Rotter B (2014) DNA fingerprinting in botany: past, present, future. Investig Genet 5:1
3. Varshney RK, Nayak SN, May GD, Jackson SA (2009) Next-generation sequencing technologies and their implications for crop genetics and breeding. Trends Biotechnol 27:522–530
4. Borevitz JO, Liang D, Plouffe D, Chang H-S, Zhu T et al (2003) Large-scale identification of single-feature polymorphisms in complex genomes. Genome Res 13:513–523
5. Barbazuk WB, Emrich SJ, Chen HD, Li L, Schnable PS (2007) SNP discovery via 454 transcriptome sequencing. Plant J 51:910–918

6. Novaes E, Drost DR, Farmerie WG, Pappas GJ, Grattapaglia D et al (2008) High-throughput gene and SNP discovery in *Eucalyptus grandis*, an uncharacterized genome. BMC Genomics 9:312
7. Garvin M, Saitoh K, Gharrett A (2010) Application of single nucleotide polymorphisms to non-model species: a technical review. Mol Ecol Resour 10:915–934
8. Huang X, Feng Q, Qian Q, Zhao Q, Wang L et al (2009) High-throughput genotyping by whole-genome resequencing. Genome Res 19:1068–1076
9. Elshire RJ, Glaubitz JC, Sun Q, Poland JA, Kawamoto K et al (2011) A robust, simple genotyping-by-sequencing (GBS) approach for high diversity species. PLoS One 6:e19379
10. Fu YB, Peterson GW (2011) Genetic diversity analysis with 454 pyrosequencing and genomic reduction confirmed the eastern and western division in the cultivated barley gene pool. Plant Genome 4:226–237
11. Poland JA, Rife TW (2012) Genotyping-by-sequencing for plant breeding and genetics. Plant Genome 5:92–102
12. Deschamps S, Llaca V, May GD (2012) Genotyping-by-sequencing in plants. Biology 1:460–483
13. Bräutigam A, Gowik U (2010) What can next generation sequencing do for you? Next generation sequencing as a valuable tool in plant research. Plant Biol 12:831–841
14. Metzker ML (2010) Sequencing technologies—the next generation. Nat Rev Genet 11:31–46
15. Altshuler D, Pollara VJ, Cowles CR, Van Etten WJ, Baldwin J et al (2000) An SNP map of the human genome generated by reduced representation shotgun sequencing. Nature 407:513–516
16. Davey JW, Hohenlohe PA, Etter PD, Boone JQ, Catchen JM et al (2011) Genome-wide genetic marker discovery and genotyping using next-generation sequencing. Nat Rev Genet 12:499–510
17. Fu YB, Cheng B, Peterson GW (2014) Genetic diversity analysis of yellow mustard (*Sinapis alba* L.) germplasm based on genotyping by sequencing. Genet Resour Crop Evol 61:579–594
18. Kim C, Guo H, Kong W, Chandnani R, Shuang L-S et al (2015) Application of genotyping by sequencing technology to a variety of crop breeding programs. Plant Sci 242:14–22
19. Baird NA, Etter PD, Atwood TS, Currey MC, Shiver AL et al (2008) Rapid SNP discovery and genetic mapping using sequenced RAD markers. PLoS One 3:e3376
20. Poland JA, Brown PJ, Sorrells ME, Jannink J-L (2012) Development of high-density genetic maps for barley and wheat using a novel two-enzyme genotyping-by-sequencing approach. PLoS One 7:e32253
21. Poland J, Endelman J, Dawson J, Rutkoski J, Wu S et al (2012) Genomic selection in wheat breeding using genotyping-by-sequencing. Plant Genome 5:103–113
22. He J, Zhao X, Laroche A, Lu Z-X, Liu H, Li Z (2014) Genotyping-by-sequencing (GBS), an ultimate marker-assisted selection (MAS) tool to accelerate plant breeding. Front Plant Sci 5:484
23. Lu F, Lipka AE, Glaubitz J, Elshire R, Cherney JH et al (2013) Switchgrass genomic diversity, ploidy, and evolution: novel insights from a network-based SNP discovery protocol. PLoS Genet 9:e1003215
24. Beissinger TM, Hirsch CN, Sekhon RS, Foerster JM, Johnson JM et al (2013) Marker density and read depth for genotyping populations using genotyping-by-sequencing. Genetics 193:1073–1081
25. Puritz JB, Matz MV, Toonen RJ, Weber JN, Bolnick DI et al (2014) Demystifying the RAD fad. Mol Ecol 23:5937–5942
26. Fu YB (2014) Genetic diversity analysis of highly incomplete SNP genotype data with imputations: an empirical assessment. G3 4:891–900
27. Jiang Z (2015) Mining next generation sequencing data: how to avoid "treasure in, error out". J Data Mining Genomics Proteomics 6:e119
28. Treangen TJ, Salzberg SL (2012) Repetitive DNA and next-generation sequencing: computational challenges and solutions. Nat Rev Genet 13:36–46
29. Clevenger J, Chavarro C, Pearl SA, Ozias-Akins P, Jackson SA (2015) Single nucleotide polymorphism identification in polyploids: a review, example, and recommendations. Mol Plant 8:831–846
30. Nevado B, Ramos-Onsins S, Perez-Enciso M (2014) Resequencing studies of non-model organisms using closely related reference genomes: optimal experimental designs and bioinformatics approaches for population genomics. Mol Ecol 23:1764–1779
31. Fu YB, Dong Y (2015) paSNPg: a GBS-based pipeline for protein-associated SNP discovery and genotyping in non-model species. J Proteomics Bioinform 8:190–194
32. Peterson GW, Dong Y, Horbach C, Fu YB (2014) Genotyping-by-sequencing for plant

genetic diversity analysis: a lab guide for SNP genotyping. Diversity 6:665–680

33. Gore MA, Chia J-M, Elshire RJ, Sun Q, Ersoz ES et al (2009) A first-generation haplotype map of maize. Science 326:1115–1117
34. Peterson BK, Weber JN, Kay EH, Fisher HS, Hoekstra HE (2012) Double digest RADseq: an inexpensive method for *de novo* SNP discovery and genotyping in model and non-model species. PLoS One 7:e37135
35. Wang S, Meyer E, McKay JK, Matz MV (2012) 2b-RAD: a simple and flexible method for genome-wide genotyping. Nat Methods 9:808–810
36. Fu YB, Peterson GW (2012) Developing genomic resources in two *Linum* species via 454 pyrosequencing and genomic reduction. Mol Ecol Resour 12:492–500
37. Parchman TL, Gompert Z, Mudge J, Schilkey FD, Benkman CW et al (2012) Genome-wide association genetics of an adaptive trait in lodgepole pine. Mol Ecol 21:2991–3005
38. De Donato M, Peters SO, Mitchell SE, Hussain T, Imumorin IG (2013) Genotyping by sequencing (GBS): a novel, efficient and cost-effective method for cattle using next-generation sequencing. PLoS One 8:e62137
39. Li H, Handsaker B, Wysoker A, Fennell T, Ruan J et al (2009) The sequence alignment/map format and SAMtools. Bioinformatics 25:2078–2079
40. Catchen JM, Amores A, Hohenlohe P, Cresko W, Postlethwait JH (2011) Stacks: building and genotyping loci de novo from short-read sequences. G3 1:171–182
41. Glaubitz JC, Casstevens TM, Lu F, Harriman J, Elshire RJ et al (2014) TASSEL-GBS: a high capacity genotyping by sequencing analysis pipeline. PLoS One 9:e90346
42. Van Orsouw NJ, Hogers RC, Janssen A, Yalcin F, Snoeijers S et al (2007) Complexity reduction of polymorphic sequences (CRoPS): a novel approach for large-scale polymorphism discovery in complex genomes. PLoS One 2:e1172
43. Toonen RJ, Puritz JB, Forsman ZH, Whitney JL, Fernandez-Silva I et al (2013) ezRAD: a simplified method for genomic genotyping in non-model organisms. Peer J 1:e203
44. Stolle E, Moritz RFA (2013) RESTseq-efficient benchtop population genomics with RESTriction Fragment SEQuencing. PLoS One 8:e63960
45. Suyama Y, Matsuki Y (2015) MIG-seq: an effective PCR-based method for genome-wide single-nucleotide polymorphism genotyping using the next-generation sequencing platform. Sci Rep 5:16963
46. Herten K, Hestand MS, Vermeesch JR, Van Houdt JK (2015) GBSX: a toolkit for experimental design and demultiplexing genotyping by sequencing experiments. BMC Bioinformatics 16:73
47. Fu YB, Peterson GW, Dong Y (2016) Increasing genome sampling and improving SNP genotyping for genotyping-by-sequencing with new combinations of restriction enzymes. G3 6:845–856
48. Munroe DJ, Harris TJR (2010) Third-generation sequencing fireworks at Marco Island. Nat Biotechnol 28:426–428
49. Kagale S, Koh C, Clarke WE, Bollina V, Parkin IA et al (2015) Analysis of genotyping-by-sequencing (GBS) Data. In: Edwards D (ed) Plant bioinformatics. Methods and protocols, vol 1374, Methods in molecular biology. Springer Science+Business Media, New York, pp 269–284
50. Eaton DAR (2014) PyRAD: assembly of *de novo* RADseq loci for phylogenetic analyses. Bioinformatics 30:1844–1849
51. Clevenger JP, Ozias-Akins P (2015) SWEEP: a tool for filtering high-quality SNPs in polyploid crops. G3 5:1797–1803
52. McCormick RF, Truong SK, Mullet JE (2015) RIG: recalibration and interrelation of genomic sequence data with the GATK. G3 5:655–665
53. Tinker NA, Bekele WA, Hattori J (2016) Haplotag: software for haplotype-based genotyping-by-sequencing analysis. G3 6:857–863
54. Cooke TF, Yee MC, Muzzio M, Sockell A, Bell R et al (2016) GBStools: a unified approach for reduced representation sequencing and genotyping. PLoS Genet 12:e1005631
55. Gordon A (2009) FASTX-Toolkit., http://hannonlab.cshl.edu/fastx_toolkit/. Accessed 21 Jan 2016
56. Chikhi R, Rizk G (2013) Space-efficient and exact De Bruijn graph representation based on a Bloom filter. Algorithms Mol Biol 8:22
57. Langmead B, Salzberg S (2012) Fast gapped-read alignment with Bowtie 2. Nat Methods 9:357–359
58. Nielsen R, Paul JS, Albrechtsen A, Song YS (2011) Genotype and SNP calling from next-generation sequencing data. Nat Rev Genet 12:443–451
59. DePristo MA, Banks E, Poplin R, Garimella KV, Maguire JR et al (2011) A framework for variation discovery and genotyping using next-generation DNA sequencing data. Nat Genet 43:491–498
60. Atwell S, Huang YS, Vilhjálmsson BJ, Willems G, Horton M et al (2010) Genome-wide

association study of 107 phenotypes in *Arabidopsis thaliana* inbred lines. Nature 465:627–631

61. Escudero M, Eaton DA, Hahn M, Hipp AL (2014) Genotyping-by-sequencing as a tool to infer phylogeny and ancestral hybridization: a case study in Carex (*Cyperaceae*). Mol Phylogenet Evol 79:359–367
62. Crossa J, Beyene Y, Kassa S, Pérez P, Hickey JM et al (2013) Genomic prediction in maize breeding populations with genotyping-by-sequencing. G3 3:1903–1926
63. Courtois B, Audebert A, Dardou A, Roques S, Ghneim-Herrera T et al (2013) Genome-wide association mapping of root traits in a japonica rice panel. PLoS One 8:e78037
64. Jarquín D, Kocak K, Posadas L, Hyma K, Jedlicka J et al (2014) Genotyping by sequencing for genomic prediction in a soybean breeding population. BMC Genomics 15:740
65. Azmach G, Gedil M, Menkir A, Spillane C (2013) Marker-trait association analysis of functional gene markers for *provitamin* A levels across diverse tropical yellow maize inbred lines. BMC Plant Biol 13:227
66. Thurber C, Ma J, Higgins R, Brown P (2013) Retrospective genomic analysis of sorghum adaptation to temperate-zone grain production. Genome Biol 14:R16
67. Hegarty M, Yadav R, Lee M, Armstead I, Sanderson R et al (2013) Genotyping by RAD sequencing enables mapping of fatty acid composition traits in perennial ryegrass (*Lolium perenne* (L.)). Plant Biotechnol J 11:572–581
68. Raman H, Raman R, Kilian A, Detering F, Carling J et al (2014) Genome-wide delineation of natural variation for pod shatter resistance in *Brassica napus*. PLoS One 9:e101673
69. Slavov GT, Nipper R, Robson P, Farrar K, Allison GG et al (2014) Genome-wide association studies and prediction of 17 traits related to phenology, biomass and cell wall composition in the energy grass *Miscanthus sinensis*. New Phytol 201:1227–1239
70. Ramstein GP, Lipka AE, Lu F, Costich DE, Cherney JH et al (2015) Genome-wide association study based on multiple imputation with low-depth sequencing data: application to biofuel traits in reed canarygrass. G3 5:891–909
71. Poland J (2016) Advances and challenges in genomic selection for disease resistance. Annu Rev Phytopathol 54:79–98
72. Tayeh N, Klein A, Le Paslier MC, Jacquin F, Houtin H et al (2015) Genomic prediction in pea: effect of marker density and training population size and composition on prediction accuracy. Front Plant Sci 6:941
73. Li X, Wei Y, Acharya A, Hansen JL, Crawford JL et al (2015) Genomic prediction of biomass yield in two selection cycles of a tetraploid alfalfa breeding population. Plant Genome 8:1–10
74. Andon M, Anderson J (2008) The oatmeal-cholesterol connection: 10 years later. Am J Lifestyle Med 2:51–57
75. Achleitner A, Tinker NA, Zechner E, Buerstmayr H (2008) Genetic diversity among oat varieties of worldwide origin and associations of AFLP markers with quantitative traits. Theor Appl Genet 117:1041–1053
76. Tinker NA, Kilian A, Wight CP, Heller-Uszynska K, Wenzl P et al (2009) New DArT markers for oat provide enhanced map coverage and global germplasm characterization. BMC Genomics 10:39
77. Newell M, Cook D, Tinker N, Jannink J-L (2011) Population structure and linkage disequilibrium in oat (*Avena sativa* L.): implications for genome-wide association studies. Theor Appl Genet 122:623–632
78. Asoro FG, Newell MA, Scott MP, Beavis WD, Jannink JL (2013) Genome-wide association study for beta-glucan concentration in elite North American oat. Crop Sci 53:542–553
79. Oliver RE, Lazo GR, Lutz JD, Rubenfield MJ, Tinker NA et al (2011) Model SNP development for complex genomes based on hexaploid oat using high-throughput 454 sequencing technology. BMC Genomics 12:77
80. Okoń S, Kowalczyk K (2012) Description of DNA analysis techniques and their application in oat (*Avena* L.) genome research. Acta Agrobot 65:3–10
81. Montilla-Bascón G, Rispail N, Sánchez-Martín J, Rubiales D, Mur LA et al (2015) Genome-wide association study for crown rust (*Puccinia coronata* f. sp. *avenae*) and powdery mildew (*Blumeria graminis* f. sp. *avenae*) resistance in an oat (*Avena sativa*) collection of commercial varieties and landraces. Front Plant Sci 6:103
82. Tinker NA, Chao S, Lazo GR, Oliver RE, Huang YF et al (2014) A SNP genotyping array for hexaploid oat (*Avena sativa* L.). Plant Genome 7. doi:10.3835/plantgenome2014.03.0010
83. Huang YF, Poland JA, Wight CP, Jackson EW, Tinker NA (2014) Using genotyping-by-sequencing (GBS) for genomic discovery in cultivated oat. PLoS One 9:e102448

84. Gutierrez-Gonzalez JJ, Tu ZJ, Garvin DF (2013) Analysis and annotation of the hexaploid oat seed transcriptome. BMC Genomics 14:471
85. Oliver RE, Tinker NA, Lazo GR, Chao S, Jellen EN et al (2013) SNP discovery and chromosome anchoring provide the first physically-anchored hexaploid oat map and reveal synteny with model species. PLoS One 8:e58068
86. Gnanesh BN, McCartney CA, Eckstein PE, Fetch JW, Menzies JG et al (2015) Genetic analysis and molecular mapping of a seedling crown rust resistance gene in oat. Theor Appl Genet 128:247–258
87. Chaffin AS, Huang Y-F, Smith S, Bekele WA, Babiker E et al (2016) A consensus map in cultivated hexaploid oat reveals conserved grass synteny with substantial sub-genome rearrangement. Plant Genome. doi:10.3835/plantgenome2015.10.0102
88. Klos KE, Huang Y-F, Bekele WA, Obert DE, Babiker E et al (2016) Population genomics related to adaptation in elite oat germplasm. Plant Genome. doi:10.3835/plantgenome2015.10.0103
89. Fu YB, Peterson GW, Williams D, Richards KW, Mitchell Fetch J (2005) Patterns of AFLP variation in a core subset of cultivated hexaploid oat germplasm. Theor Appl Genet 111:530–539
90. Fu YB, Williams D (2008) AFLP variation in 25 *Avena* species. Theor Appl Genet 117:333–342
91. Sanz MJ, Jellen EN, Loarce Y, Irigoyen ML, Ferrer E, Fominaya A (2010) A new chromosome nomenclature system for oat (*Avena sativa* L. and *A. byzantina* C. Koch) based on FISH analysis of monosomic lines. Theor Appl Genet 121:1541–1552
92. Andrews KR, Hohenlohe PA, Miller MR, Hand BK, Seeb JE et al (2014) Trade-offs and utility of alternative RADseq methods: reply to Puritz et al. Mol Ecol 23: 5943–5946
93. Cohen L, Shen S, Efstathiadis E (2015) De novo transcriptome assembly. In: Brown SM (ed) Next-generation DNA sequencing informatics, 2nd edn. CSHL Press, New York, pp 155–168
94. Truong HT, Ramos AM, Yalcin F, de Ruiter M, van der Poel HJ et al (2012) Sequence-based genotyping for marker discovery and co-dominant scoring in germplasm and populations. PLoS One 7:e37565
95. Monson-Miller J, Sanchez-Mendez DC, Fass J, Henry IM, Tai TH et al (2012) Reference genome-independent assessment of mutation density using restriction enzyme-phased sequencing. BMC Genomics 13:72
96. Pukk L, Ahmad F, Hasan S, Kisand V, Gross R et al (2015) Less is more: extreme genome complexity reduction with ddRAD using Ion Torrent semiconductor technology. Mol Ecol Resour 15:1145–1152
97. Sonah H, Bastien M, Iquira E, Tardivel A, Légaré G et al (2013) An improved genotyping by sequencing (GBS) approach offering increased versatility and efficiency of SNP discovery and genotyping. PLoS One 8:e54603
98. Fu X, Dou J, Mao J, Su H, Jiao W et al (2013) RADtyping: an integrated package for accurate *de novo* codominant and dominant RAD genotyping in mapping populations. PLoS One 8:e79960
99. Puritz JB, Hollenbeck CM, Gold JR (2014) dDocent: a RADseq, variant-calling pipeline designed for population genomics of non-model organisms. Peer J 2:e431
100. Sovic MG, Fries AC, Gibbs HL (2015) AftrRAD: a pipeline for accurate and efficient *de novo* assembly of RADseq data. Mol Ecol Resour 15:1163–1171

Chapter 14

Genome-Wide Association Analysis Using R

Julio Isidro-Sánchez, Deniz Akdemir, and Gracia Montilla-Bascón

Abstract

This chapter provides a practical overview of the statistical analysis using R [1] and genotype by sequencing (GBS) markers for genome-wide association studies (GWAS) in oats. Statistical analysis is performed by R package rrBLUP [2] and issues associated with the analysis are addressed along with the R code. The ultimate aim of this chapter is to provide a practical guideline to do GWAS analysis using R, rather than describe the theory in depth. For more details about the subject, readers are referred to the excellent resource book in GWAS [3]. A basic programming experience in R is assumed.

Key words GWAS, Population structure, Linkage disequilibrium, False discovery rate, Bonferroni correction

1 Introduction

The central goal of GWAS is to identify casual mutations that have an effect on a phenotype (any aspect of an organism that can be measured). A casual mutation is a position in the genome where an experimental manipulation of the DNA produces an effect on the phenotype on average. From a statistical point of view, a casual mutation occurs when $\mathrm{Cov}(Y,X) \neq 0$ where Y are the value of the phenotypes and X the value of the genotypes.

Genome-wide association analyses are aimed for detecting variants at genomic loci that are associated with complex traits in the population and, in particular, at detecting associations between common single-nucleotide polymorphisms (SNPs) and common diseases. Markers that are significantly associated with the phenotype are presumed to be in linkage disequilibrium (LD) with putative Quantitative Trait Loci (QTL). The goal of GWAS, is to test for association between the frequency of each of hundreds of thousands of common variants and a given phenotype, that exceed a

Electronic Supplementary Material: The online version of this chapter (doi: 10.1007/978-1-4939-6682-0_14) contains supplementary material, which is available to authorized users.

Sebastian Gasparis (ed.), *Oat: Methods and Protocols*, Methods in Molecular Biology, vol. 1536,
DOI 10.1007/978-1-4939-6682-0_14,

conservative genome-wide threshold for association and then test these for evidence of replication. High statistical power, low probability of Type I error, use of covariates, and high resolution are the keys for success in GWAS.

From a multiple regression model, with a continuous phenotype Y, the rigorous formulation of GWAS analysis is

$$y = \beta_u + X_a\beta_a + X_d\beta_d + \varepsilon$$

where β_u is the mean, X_a and X_d are random markers variables with values of $A_1A_1 = -1, A_1A_2 = 0$ and $A_2A_2 = 1$ for additive *Xa* and $A_1A_1 = -1, A_1A_2 = 1$ and $A_2A_2 = -1$ for dominance X_d, β_a and β_d are associated with X_a and X_d respectively, and ε is the random error.

In GWAS we are testing for every marker the following hypothesis test:

$$\text{Ho}: \beta_a = 0 \cap \beta_d = 0$$
$$\text{Ha}: \beta_a \neq 0 \cap \beta_d \neq 0$$

In this chapter, the GWAS analysis is carried out using the R package rrBLUP (https://cran.r-project.org/web/packages/rrBLUP/rrBLUP.pdf). Within the package the *A.mat* and *GWAS* functions are used to perform the analysis based on the mixed model described by Yu et al. [4] as follow.

$$y = X\beta + Zg + S\tau + \varepsilon$$

where *y* is a vector of phenotype observations, β is a vector of fixed effects, g models the genetic background of each line as a random effect with $\text{Var}[g] = K\sigma^2$, τ is a vector of additive SNP effects as a fixed effect, and ε is a vector of residual effects with $\text{Var}[\varepsilon] = I\sigma_\varepsilon^2$.

It is important to point out that without additional evidence (experimental manipulation of the DNA) that the statistical models cannot inference with absolute certainty to identify a causal mutation. Genome-wide association studies are a starting point for additional experimental work.

2 Material

Our genetic system have two components that can be measured, the phenotypic and the genotypic component. The simulated data sets used here are an example modified from wheat dataset; with 932 phenotypic yield data observations, and 329 genotypic data with 3629 markers, which are mapped. Data can be downloaded from book repository (The data files "GWAS_Data.RData" and "GWAS_Script.R" can be downloaded from the link: 10.1007/978-1-4939-6682-0_14).

The analysis of GWAS is performed using genotype by sequencing (GBS), which is characterized for having many missing data (http://cbsu.tc.cornell.edu/lab/doc/GBS_Method_Overview1.pdf). Genotype by sequencing data normally generate hundred of thousands of markers. For the simplicity of the analysis, a subset of the GBS data has been selected in this study.

Although, phenotypes are considered following a normal probability model, there are a broad class of models that can apply to continuous and discrete phenotypes analysis [5]. The R software is free and open-source statistical software that can be downloaded for Windows, Mac OS X, or Linux from https://www.r-project.org/. From the left side of the website click on CRAN (Comprehensive R Archive Network), select the appropriate CRAN Mirror, and select the appropriate operating system and install it into your computer following on-screen prompts for installation. Once in your computer, rrBLUP package needs to be installed by using the next command.

>install.packages("rrBLUP") [1, 2]

3 Methods

3.1 Basic Guideline to Perform a GWAS Study

Main guidelines to do a GWAS study.

1. Read phenotypes and check the assumptions of the models.
   (a) Check outliers.
   (b) Normal distribution of errors. Possible transformation of the data.
2. Read genotypes and filter for:
   (a) Markers with a proportion of missing data more than a particular threshold set by the researcher.
   (b) Individuals with a high proportion of missing data.
   (c) Individuals with a high proportion of heterozygous.
   (d) Remove genotypes with a minor allele frequency (MAF) less than 5 %.
   (e) Remove genotypes that fail a Hardy-Weinberg test of equilibrium (Normally, use a conservative p-value cut-off of $<10^{-5}$).
3. Imputation of the genotype file, and performing the Kinship matrix.
4. Look for population structure effects.
5. Match phenotypes and genotypes.
6. Perform GWAS function from rrBLUP with and without population structure effects and Kinship matrix.
7. Manhattan and Q-Q plot graphs.
8. Interpretation and Validation.

3.2 Analysis GWAS Study in R

1. Read phenotypes and check the assumptions of the models.

```
pheno <- read.csv("phenoat.csv",header=T);
dim(pheno)

## [1] 932 3

head(pheno) ## GID ENV Yield

## 1 Oat179 Env1 6317.606
## 2 Oat130 Env2 6335.475
## 3 Oat303 Env4 7259.274
## 4 Oat270 Env1 6916.124
## 5 Oat202 Env4 6845.943
## 6 Oat233 Env3 5750.001

str(pheno)

## 'data.frame': 932 obs. of 3 variables:
## $ GID : Factor w/ 330 levels "Oat1","Oat10",..:
89 36 228 191 116 150 205 25 153 264 …
## $ ENV : Factor w/ 4 levels "Env1","Env2",..:
1 2 4 1 4 3 2 2 3 2 …
## $ Yield: num 6318 6335 7259 6916 6846 …
```

The dataset contains **932** yield observations measured in four different environments. The structure function in R indicates that GID and ENV are factors, which is the correct class vector needed for the analysis. If they were not factors, you could change the class using "as.factor()" function in R.

Data showed a continuous distribution (Fig. 1), therefore we could make inference using regression models. Some assumptions have to be met in order to make inference. Observations have to be independent and identically distributed and errors (and therefore the phenotypes) have to follow a normal distribution. Data seems to meet the normality assumption (Fig. 1), although few outliers can be observed.

```
hist(pheno$Yield,xlab="Yield",main="Histogram Yield")

shapiro.test(pheno$Yield)

##
## Shapiro-Wilk normality test
##
## data: pheno$Yield
## W = 0.99392, p-value = 0.0007821
```

The code above checks the normality of the phenotypes. In general, if phenotypes are normal the residuals will be also normal. The Shapiro–Wilk test for normality indicates that data are not normal. The solution for this problem is to check for outliers and transform data in order to get the normality of the phenotypes.

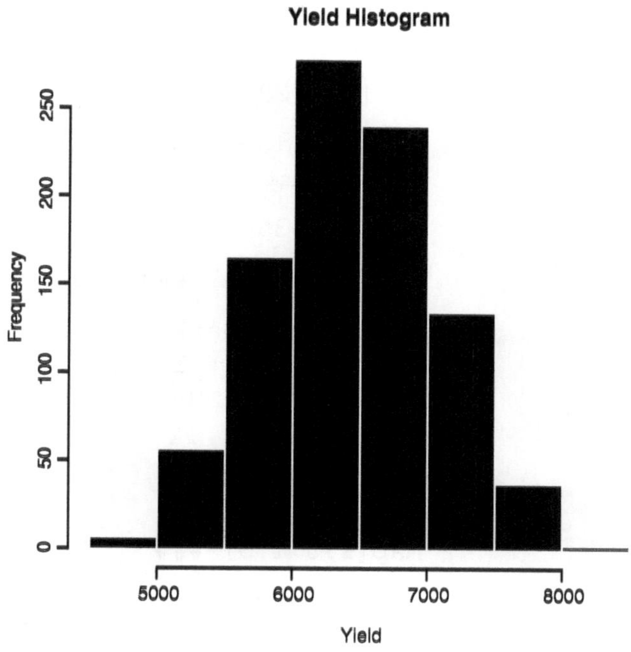

Fig. 1 Histogram of phenotypic data

```
boxplot.yield<- boxplot(pheno$Yield)
outliers <- boxplot.yield$out; outliers
## [1] 8979.334 3944.849 3947.001 3948.641 4576.506
3942.469 8166.286
## [8] 4616.835 4672.310 8920.055

pheno <- pheno[-which(pheno$Yield%in%outliers),]
shapiro.test(pheno$Yield)

##
## Shapiro-Wilk normality test
##
## data: pheno$Yield
## W = 0.99719, p-value = 0.1093

pheno <- na.omit(pheno)
```

The code above indicates that there are ten outliers on the yield phenotypic data. After removing the outliers, we cannot reject the Shapiro–Wilk normality test indicating that yield data are now normal. Last line in code is to eliminate any possible missing data (NA).

(*see* **Note 1**).

2. Read genotypes and filtering.

Genotypes and map information are read into the R working directory. Genotypic data contains 329 lines and 3629 markers. The map file contains 3354 markers and three variables.

```
geno <- read.csv("genoat.csv",header=T,row.names = 1);
dim(geno)

## [1] 329 3629

map <- read.csv("mapoat.csv",header=T,stringsAsFactors
=F,row.names=1); dim(map)

## [1] 3354 3

geno[1:5,1:5] ### View genotypic data.

## Marker1 Marker2 Marker3 Marker4 Marker5
## Oat1  NA   1   1  1   1
## Oat2  -1  NA   1  1   1
## Oat3  -1   1   1  1  NA
## Oat4  -1  NA  -1  1   1
## Oat5  -1  NA  -1  1   1

map[1:5,1:3] ###

## Markers chrom loc
## 1  Marker607  1A 16730
## 2 Marker2900  1A 16730
## 3 Marker1316  1A 25338
## 4 Marker2297  1A 26595
## 5 Marker1895  1A 27071
```

The next step is to filter the genotypic data. Filtering conditions will depend on researcher criteria. The code below represent a simple function to remove individuals and markers that does not met the criteria establish by the researcher. The function will remove individuals with more than a certain percentage of missing data, markers with a greater proportion of a threshold missing percentage, and also markers with a high proportion of heterozygous calls.

```
filter.fun <- function(geno,IM,MM,H){
  #Remove individuals with more than a certain %
  missing data  individual.missing <- apply(geno,1,
  function(x){
    return(length(which(is.na(x)))/ncol(geno))
  })

  #Remove markers with certain % missing data
  marker.missing <- apply(geno,2,function(x)
  {return(length(which(is.na(x)))/nrow(geno))

  })
  length(which(marker.missing>0.6))
  #Remove individuals with high heterozygous calls.
  heteroz <- apply(geno,1,function(x){
    return(length(which(x==0))/length(!is.na(x)))
  })
```

```
    filter1 <- geno[which(individual.missing<IM),
which(marker.missing<MM)]
      filter2 <- filter1[,(heteroz<H)]
      return(filter2)
   }
```

(*see* **Note 2**).

```
geno.filtered <- filter.fun(geno[,1:3629],0.4,0.60,0.02)
```

geno.filtered will be composed by those genotypes with less than 40% missing data, markers with less than 60% missing data and individuals with less than 2% of heterozygous calls. After filtering one individual have been removed and 361 markers.

```
geno.filtered[1:5,1:5];dim(geno.filtered)

##    Marker1 Marker2 Marker3 Marker4 Marker5
## Oat1    NA       1       1       1       1
## Oat2    -1      NA       1       1       1
## Oat3    -1       1       1       1      NA
## Oat4    -1      NA      -1       1       1
## Oat5    -1      NA      -1       1       1

## [1] 328 3268
```

The lower the minor allele frequency (MAF) the lower the statistical power, because MAF increases the variance of the phenotypes associated with the MAF alleles. Therefore, it is necessary to filter them and the most standard way is to eliminate markers with less than 5% of MAF. Minor allele frequency will be removed using the A.mat function within the rrBLUP package.

3. Imputation of the genotype file, and performing the Kinship matrix.

   The main idea behind imputation is to predict (or 'impute') the missing data based upon the observed data. Imputation is now routinely used to facility genotyped studies by increasing the power of the analysis. Here, "A.mat" function from rrBLUP package is used for imputation. A.mat has two options for imputation (https://cran.r-project.org/web/packages/rrBLUP/rrBLUP.pdf). One is to replace missing data with the population mean for that marker, or using an expectation maximization (EM) algorithm based on the multivariate normal distribution [6].

```
library(rrBLUP)
Imputation    <-    A.mat(geno.filtered,impute.method="EM",
return.imputed=T,min.MAF=0.05)

## [1] "A.mat converging:"
## [1] 0.0431
## [1] 0.00647
```

```
K.mat <- Imputation$A ; dim(K.mat) ### KINSHIP matrix

## [1] 328 328

geno.gwas <- Imputation$imputed; dim(geno.gwas) #NEW geno
data.

## [1] 328 1962

geno.gwas[1:5,1:5]## view geno

##         Marker1 Marker3 Marker4 Marker5 Marker7
## Oat1 -1.44066       1       1 1.0000000      -1
## Oat2 -1.00000       1       1 1.0000000      -1
## Oat3 -1.00000       1       1 0.6753843      -1
## Oat4 -1.00000      -1       1 1.0000000      -1
## Oat5 -1.00000      -1       1 1.0000000      -1

K.mat[1:5,1:5]## view Kinship

##           Oat1         Oat2       Oat3        Oat4         Oat5
## Oat1 1.8191561  0.220454719 0.1009423 0.15218203 0.177660657
## Oat2 0.2204547  2.111943356 0.1266988 0.02978199 -0.007214392
## Oat3 0.1009423  0.126698834 1.8000414 -0.12678093 -0.126589635
## Oat4 0.1521820  0.029781989 -0.1267809 1.87520005 1.803271696
## Oat5 0.1776607 -0.007214392 -0.1265896 1.80327170 1.873595678
```

A.mat estimate the relationship matrix using the EM algorithm, and also return the genotypic information. In this function, we also removed the MAF markes having at the end of the filtering 328 genotypes and 1962 markers.

4. Look for population structure effects.

One crucial step in GWAS analysis is to study the population structure (PS). The main reason to perform this study is that, as a consequence of having different population genetic histories, distinct subpopulations could have differences in allele frequencies for many polymorphisms throughout the genome. If the populations have different overall values for the phenotype, any polymorphisms that differ in frequency between the two populations are associated with the phenotype even though they are not casual or in strong linkage disequilibrium with casual polymorphisms [7–9]. Principal component analysis (PCA) on genotypic data is used to visualize the structure of our populations using the function "*svd()*" in R.

Population structure accounted by PCA is limited to correcting for spurious associations on a global level of genetic variation. Thereby, PS does not adequately capture the relatedness between individuals, and this relationship between genotypes (K, kinship matrix) needs also be taking into account on the analysis. Not taking into account of PS, K as well as a potential confounding between the phenotype and the genotype effects, could lead to unrealistic assessments in GWAS analysis.

```
geno.scale <- scale(geno.gwas,center=T,scale=F) # Data needs to
be center.
Svdgeno <- svd(geno.scale)
PCA <- geno.scale%*%svdgeno$v #Principal components
colnames(PCA) <- paste("PCA",1:ncol(PCA),sep="")
PCA[1:5,1:5]
```

```
##              PCA1        PCA2      PCA3       PCA4        PCA5
## Oat1  -6.976549   -9.824339  5.290272  -1.184019  5.00415366
## Oat2  -8.745900  -10.545057  4.730541  -8.834242 -0.02504027
## Oat3  -6.292199    3.412671  4.774038   6.657838  8.93925754
## Oat4 -19.524147    7.689386 -7.963162 -13.644082 -6.19429306
## Oat5 -19.334593    8.011715 -7.624697 -13.033030 -6.53379527
```

Principal component analysis is calculated multiplying the scaled genotypes times the eigenvectors. The total number of PCA is equal to the total number of lines on the genotypic file. The total genetic variance explained by the PCA can be seen on Fig. 2.

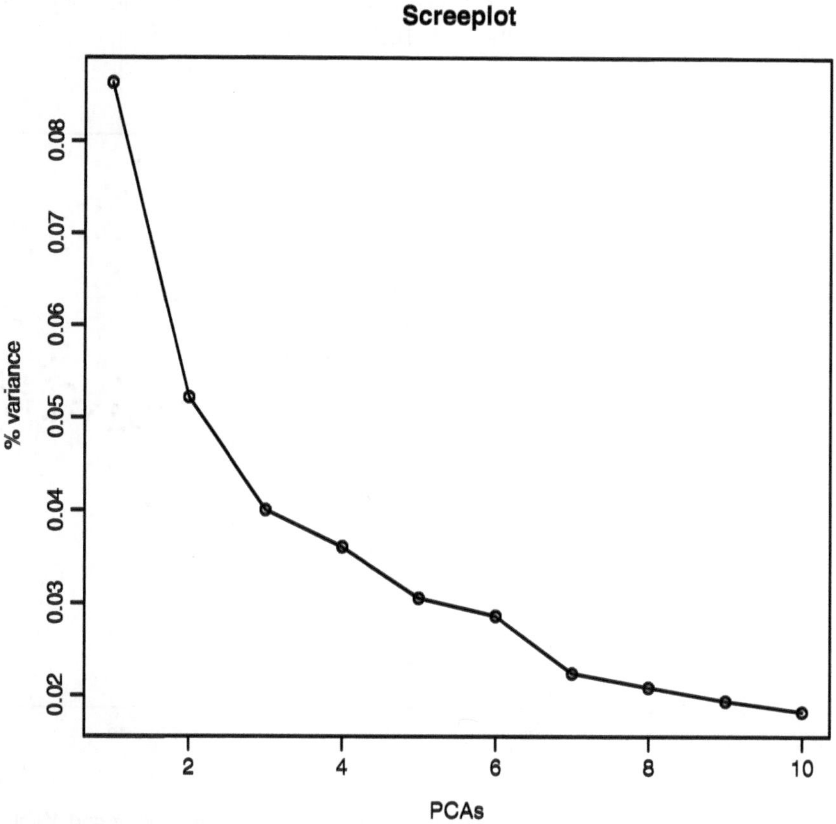

Fig. 2 Screeplot representing the percentage of the variance explained by the principal components

```
plot(round((svdgeno$d)^2/sum((svdgeno$d)^2)
,d=7)[1:10],type="o",main="Screeplot",xlab="PC
As",ylab="% variance")
```

The screeplot indicates that the first two PC explained around 14% of the genetic variance, indicating that population structure effects are mild in this dataset. Figure 3 shows the first two PC.

```
PCA1 <- 100*round((svdgeno$d[1])^2/sum((svdgeno$
d)^2),d=3); PCA1
## [1] 8.6
PCA2 <- 100*round((svdgeno$d[2])^2/sum((svdgeno$
d)^2),d=3); PCA2
## [1] 5.2
```

Depending on the genotype and breeder information, a clustering analysis we could highlight different clusters on the graph. The following script shows how to make clusters and plot the first tow clusters on the PC graph (Fig. 4).

```
Eucl <- dist(geno.gwas) # Euclinean distance
Fit <- hclust(Eucl,method="ward") # Ward crite-
rion makes clusters with same size.

## The "ward" method has been renamed to "ward.D";
note new "ward.D2"
```

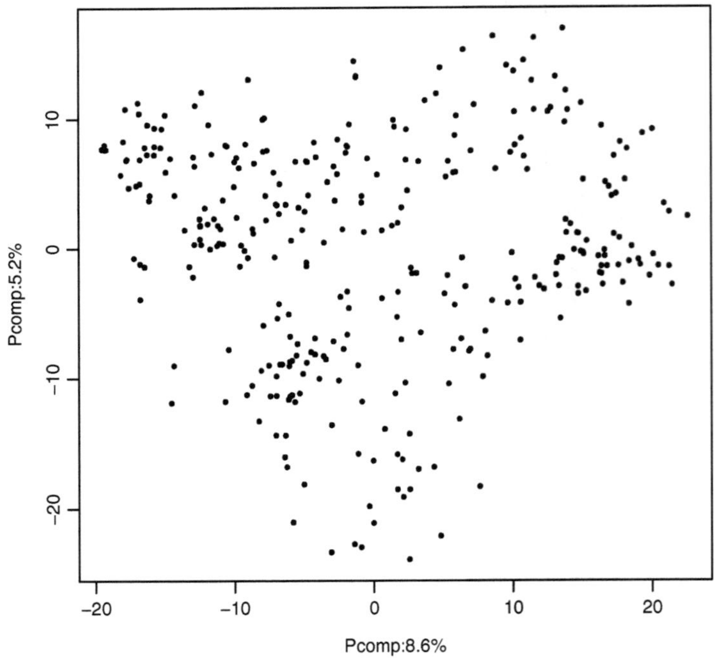

Fig. 3 First and second principal component analysis. *X* and *Y* values represent the proportion of variance explained by each PC

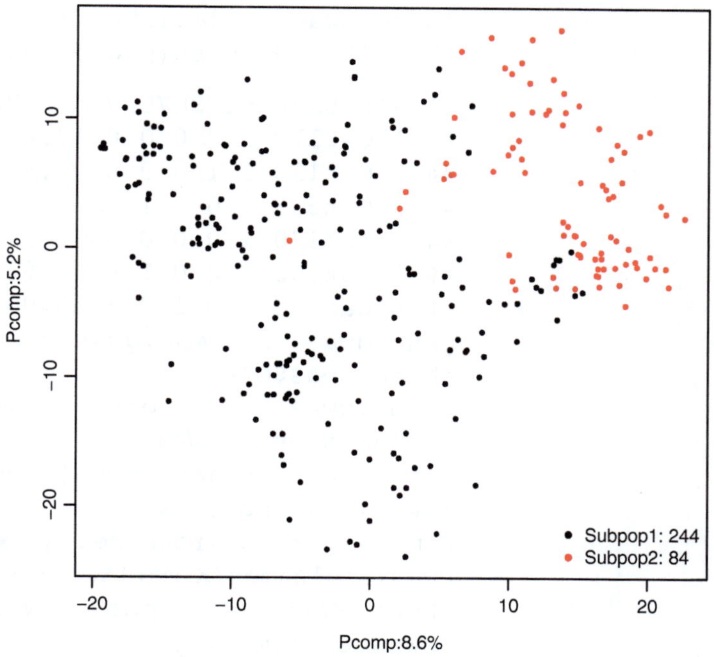

Fig. 4 Principal component analysis plots with the clustering information. *Red* and *black circles* represent two distinct subpopulations

```
groups2 <- cutree(fit,k=2) # Selecting two clusters.
table(groups2) #Number of individuals per cluster.
```

```
## groups2
## 1 2
## 244 84
```

```
plot(PCA[,1],PCA[ ,2],xlab=paste("Pcomp:",PCA1,"%",sep=""),ylab=paste("Pcomp:",PCA2,"%",sep=""),pch=20,cex=0.7,col=groups2)
```

5. Match phenotypes and genotypes.

   The following code prepares the files to be used on GWAS function. In order to do the analysis the same genotypes needs to be on phenotype and genotypes files. A matching function between files needs to be performed.

```
pheno=pheno[pheno$GID%in%rownames(geno.gwas),]
pheno$GID<-factor(as.character(pheno$GID),
levels=rownames(geno.gwas)) #to assure same lev-
els on both files
##Creating file for GWAS function from rrBLUP
package X<-model.matrix(~-1+ENV, data=pheno)
```

```
pheno.gwas <- data.frame(GID=pheno$GID,X,Yield=
pheno$Yield) head(pheno.gwas)

## GID ENVEnv1 ENVEnv2 ENVEnv3 ENVEnv4 Yield
## 1 Oat179 1 0 0 0 6317.606
## 2 Oat130 0 1 0 0 6335.475
## 3 Oat303 0 0 0 1 7259.274
## 4 Oat270 1 0 0 0 6916.124
## 5 Oat202 0 0 0 1 6845.943
## 6 Oat233 0 0 1 0 5750.001
geno.gwas <- geno.gwas[rownames(geno.gwas)%in%
pheno.gwas$GID,]
pheno.gwas   <-   pheno.gwas[pheno.gwas$GID%in%
rownames(geno.gwas),]
geno.gwas <- geno.gwas[rownames(geno.gwas)%in%
rownames(K.mat),]
K.mat <- K.mat[rownames(K.mat)%in%rownames(geno.
gwas),colnames(K.mat)%in%rownames(geno.gwas)]
pheno.gwas   <-   pheno.gwas[pheno.gwas$GID%in%
rownames(K.mat),]
```

Likewise, same information needs to be between genotypes and map information. The next code matches both files.

```
geno.gwas  <-geno.gwas[,match(map$Markers,colnam
es(geno.gwas))] head(map)

##        Markers  chrom     loc
## 1    Marker607     1A   16730
## 2   Marker2900     1A   16730
## 3   Marker1316     1A   25338
## 4   Marker2297     1A   26595
## 5   Marker1895     1A   27071
## 6   Marker2902     1A   27071

geno.gwas <- geno.gwas[,colnames(geno.gwas)%in%map
$Markers]
map <- map[map$Markers%in%colnames(geno.gwas),]
geno.gwas2<- data.frame(mark=colnames(geno.gwas),
chr=map$chrom,loc=map$loc,t(geno.gwas))
dim(geno.gwas2)

## [1] 1759 331

colnames(geno.gwas2)[4:ncol(geno.gwas2)]       <-
rownames(geno.gwas)
```

This is the final view of the three different files needed for the GWAS analysis in rrBLUP package.

```
head(pheno.gwas)

## GID ENVEnv1 ENVEnv2 ENVEnv3 ENVEnv4 Yield
## 1 Oat179 1 0 0 0 6317.606
```

```
## 2 Oat130 0 1 0 0 6335.475
## 3 Oat303 0 0 0 1 7259.274
## 4 Oat270 1 0 0 0 6916.124
## 5 Oat202 0 0 0 1 6845.943
## 6 Oat233 0 0 1 0 5750.001
```

```
geno.gwas2[1:6,1:6]
```

```
## mark chr loc Oat1 Oat2 Oat3
## Marker2297 Marker2297 1A 26595 -1 -1.0000000
-1
## Marker3125 Marker3125 1A 35232 1 -1.0000000
-1
## Marker2100 Marker2100 1A 35653 -1 1.0000000 1
## Marker1797 Marker1797 1A 51943 1 -0.4259651
-1
## Marker3191 Marker3191 1A 51943 -1 1.0000000 1
## Marker1403 Marker1403 1A 56393 -1 1.0000000 1
```

```
K.mat[1:6,1:6]
```

```
## Oat1 Oat2 Oat3 Oat4 Oat5
## Oat1 1.8191561 0.220454719 0.1009423 0.15218203
0.177660657
## Oat2 0.2204547 2.111943356 0.1266988 0.02978199
-0.007214392
## Oat3 0.1009423 0.126698834 1.8000414
-0.12678093 -0.126589635
## Oat4 0.1521820 0.029781989 -0.1267809
1.87520005 1.803271696
## Oat5 0.1776607 -0.007214392 -0.1265896
1.80327170 1.873595678
## Oat6 -0.1810759 -0.230823204 0.1734623
-0.40927660 -0.357916219
```

6. Perform GWAS function from rrBLUP with and without population structure effects and Kinship matrix.

   A statistically significant association between a genotypic marker and a particular trait is considered to be a proof of linkage between the phenotype and a casual locus. Generally, PS leads to spurious associations between markers and a trait, so that a statistical approach must account for PS [10].

   In this analysis, four different statistical models were performed.

   (a) Naïve model without controlling for PS or family relatedness (gwasresults).
   (b) Controlling for PS effects (Q model, gwasresults2).
   (c) Controlling just for relatedness (K model, gwasresults3).
   (d) Controlling for both Q and K effects (Q+K model, gwasresults4).

```
## gwasresults<-GWAS(pheno.gwas,geno.gwas2, fixed=
colnames(pheno.gwas)[2:5], K=NULL, plot=T,n.PC=0)

##gwasresults2<-GWAS(pheno.gwas,geno.gwas2,
fixed=colnames(pheno.gwas)[2:5], K=NULL, plot=T,n.
PC=6)

##gwasresults3<-GWAS(pheno.gwas,geno.gwas2,
fixed=colnames(pheno.gwas)[2:5], K=K.mat, plot=T,n.
PC=0)

##gwasresults4<-GWAS(pheno.gwas,geno.gwas2,
fixed=colnames(pheno.gwas)[2:5], K=K.mat, plot=T,n.
PC = 6)
```

7. Manhattan and Q-Q plot graphs.

   With the aim to provide more scrip for R learning readers, we report here the Q-Q and Manhattan plot using R. The readers must know that GWAS function has an option to plot both graphs (plot = T).

   Once your GWAS analysis is complete, the minimum set of analysis one should undertake when presenting your GWAS analysis are the Quantile-Quantile (Q-Q) and the Manhattan plot.

   The Q-Q plot is a visual diagnostic tool that can identify the effects of accounted for population structure, relatedness and other issues in GWAS analysis. A Q-Q plot is use to detect a difference between the distribution of *p*-values that we observe in GWAS compared to the *p*-values we would have expected to produce if the null hypothesis (no association) was true for every marker we tested. How the Q-Q plot differs from this null expectation will provide us with a considerable amount of information about whether or not something is amiss with our statistical analysis. In Fig. 5, the Q-Q plots for all models are presented.

   As we checked on **step 4**, the PS effects were not strong, which indicates that the effect on the analysis will not be really significant. That is the reason why although including PS effects the change on the Q-Q plot were not substantial. We used *"n.PC = 6"* to account for PS because it will account for most of the variance explained by the genotypes.

   In general, Q-Q plots that do not look like the null expectation indicate that something is wrong with our statistical analysis of our GWAs data (accounted for factors, large number of genotyping errors, violations of the assumptions of our statistical tests, etc.). In such cases, we cannot draw any conclusions concerning the detection of casual polymorphism.

   A typical report visualization of GWAS of SNPs is the Manhattan plot of significance against chromosomal location (Fig. 6). In a GWAS, the goal is that any marker for which we have rejected the null hypothesis, to be in LD with a casual polymorphism. One of the reasons why a test we have rejected the null is

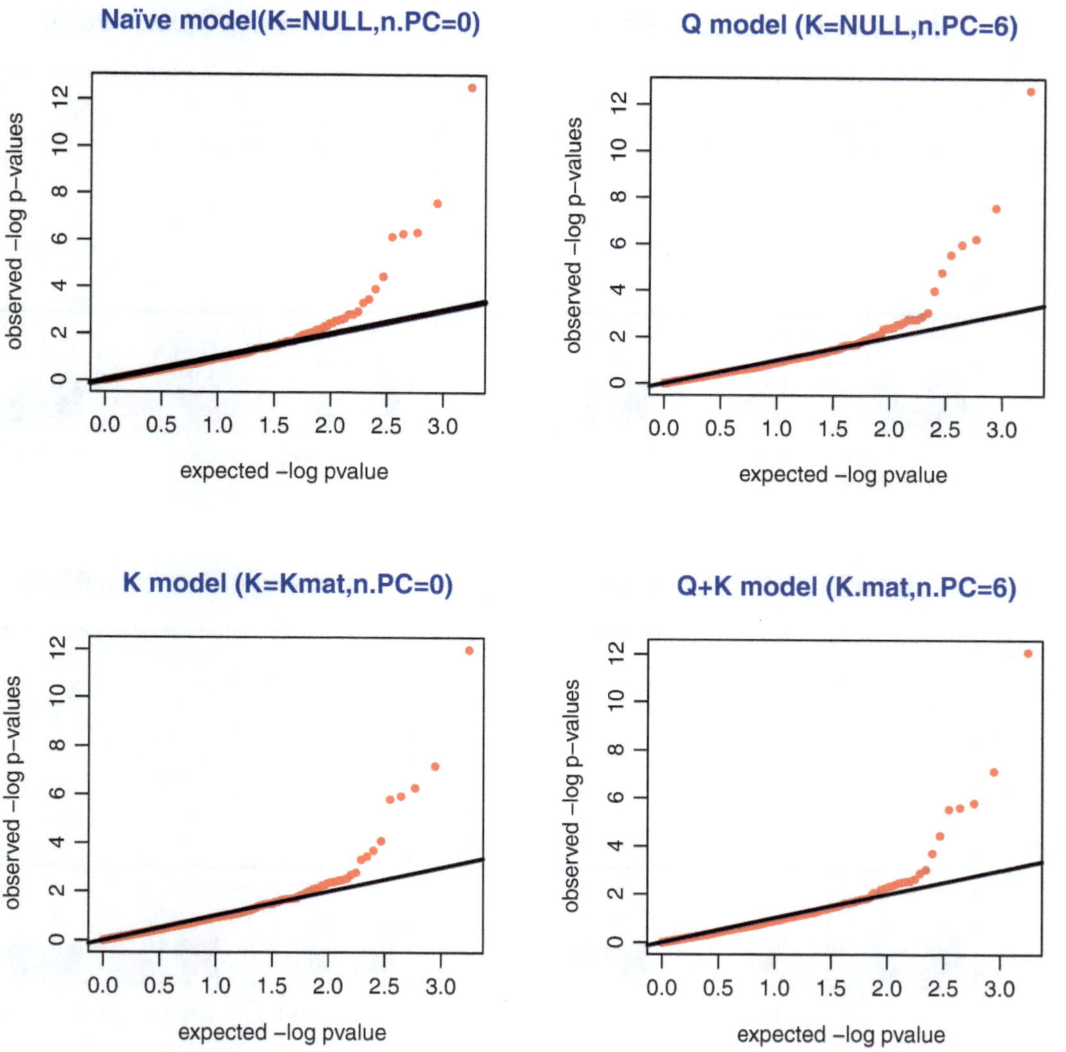

Fig. 5 Q-Q plots of the different models. The Q-Q plots for all models seem to follow the 1:1 ratio expectation. When accounting for PS on the analysis (Q model), more signal is captured by the model, as the *red line* follows closer the 1:1 expected ratio line, and in this sense the K model seems to fit better than Q model. The Q + K model did not improve significantly the analysis. This indicates the kinship matrix is enough to account for *solid line*. This makes sense since PS effects were mild as the PCA analysis shows on Fig. 3

not in LD with the casual polymorphism is the Type I Error (the probability of rejecting the null hypothesis when the null is true) from the multiple testing problem. Analysis needs to be corrected by Type I error because it would be very costly to track wrong significant casual polymorphisms. Type I error rate can be controlled by setting alpha lower. This is a trade-off: the lower the Type I error is set, the lower the power of our hypothesis. In this analysis, Bonferroni correction and False Discovered Rate (FDR) are presented. Those markers above the threshold are considered to be in LD with the casual polymorphism (Fig. 6).

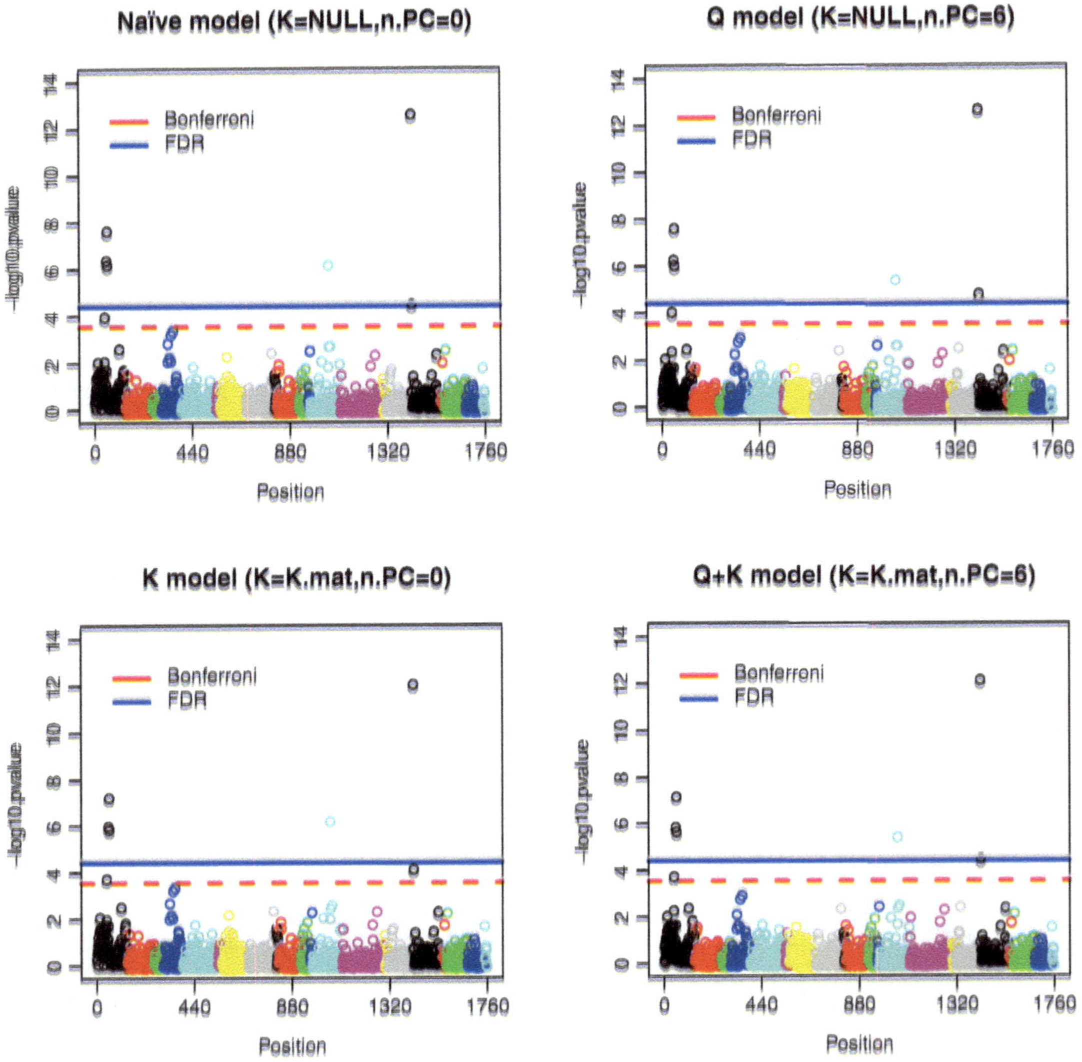

Fig. 6 Manhattan plots of the different statistical models. Threshold are indicated with *lines* in *red* (Bonferroni correction) and *blue* (False Discovery Rate). Chromosome positions are drawn with different colors on *x* axis)

(*see* **Note 3**).

The significant hits from the Manhattan plots on Q+K model are:

```
alpha_bonferroni=-log10(0.05/length(gwasresults
$Yield)) ###This is Bonferroni correcton
alpha_FDR_Yield<--log10(FDR(10^(-gwasresults$Yield),
0.05))## This is FDR cut off
```

```
which(gwasresults4$Yield>alpha_bonferroni)
```

```
## [1] 53 56 57 1054 1427
```

```
which(gwasresults4$Yield>alpha_FDR_Yield)
```

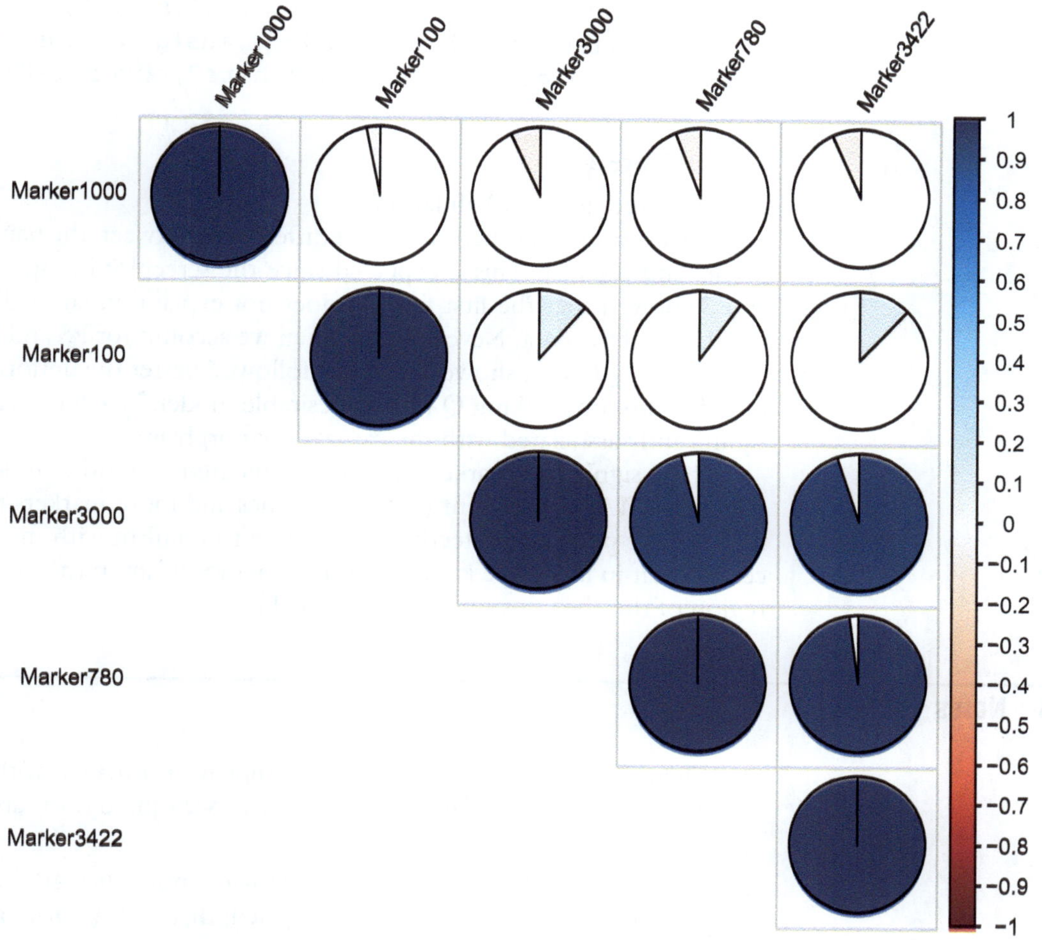

Fig. 7 Correlation among significant hits on Q + K model. High correlations are expressed with high intensity *blue* or *red color*

```
## [1]   45   53   56   57 1054 1427 1428
```

These are the significant *p*-values. We expect a set of polymorphism that are close to each other and high correlated in the genome to the causal polymorphism. Figure 7 shows the correlation between the significant markers. The highest correlation appears to be between SNPs closer to the casual mutation, since there are in high LD with each other and therefore with a putative QTL.

```
library(corrplot)
markers.gwasresults4.bonf<-geno.gwas[,c(53,56,57,
1054,1427)]
corr_sign  <-  cor(markers.gwasresults4.bonf,
use="complete.obs") par(oma=c(2,2,4,2))
corrplot(corr_sign,order="hclust",method="pie",
tl.pos="lt", type="upper",
```

```
        tl.col="black", tl.cex=0.8, tl.srt=55,
        sig.level=0.90,cl.length=21,insig = "blank")
mtext("Correlation Significant hits",outer=TRUE,
line=1)
```

(*see* **Note 4**).

8. Interpretation and Validation.

   Results show that there are slight differences between the naïve model and the rest. This is expected since the screeplot in Figs. 2 and 3 showed that the first few PC does not explain much of the variability of the data. Nevertheless, when we account for PS and K matrix, the Q-Q plot showed that data followed better the uniform null distribution, making Q + K the desirable model. Five loci were significantly associated with the casual polymorphism:

   These significant markers should be validated in another independent GWAS experiment (with more lines and more markers to increase the power of detection) to be able to confirm with more certain that in fact there is a true casual polymorphism in the same region of the chromosomes and more QTL.

4 Notes

1. When data are not normal, you can improve normality of the original phenotypic data using logit, inverse, square root, arcsine, and log transformation methods.
2. You can check how many individuals with more that 40% of missing date will be removed by running: length(which(individual.missing > 0.40).
3. Bonferroni correction is very conservative in comparison with FDR. A Bonferroni correction of alpha = 0.01 decrease the power of detection considerably. The choice between Bonferroni and FDR depends on researcher.
4. This does not indicate that we found the casual polymorphism, but it is a food approximation of the region where the casual polymorphism might be. Information from Q-Q plot guide to use the K or Q + K model as the guideline to show the Manhattan plots results.

References

1. R Core Team (2013) R: a language and environment for statistical computing. R Foundation for Statistical Computing, Vienna, Austria, http://www.R-project.org/
2. Endelman JB (2011) Ridge regression and other kernels for genomic selection with R package rrBLUP. Plant Genome 4(3):250–255
3. Gondro C, Van der Werf J, Hayes B (eds) (2013) Genome-wide association studies and genomic prediction. Humana, New York
4. Yu J, Pressoir G, Briggs WH, Bi IV, Yamasaki M, Doebley JF, McMullen MD, Gaut BS, Nielsen DM, Holland JB, Kresovich S, Buckler ES (2006) A unified mixed-model method for

association mapping that accounts for multiple levels of relatedness. Nat Genet 38:203–208
5. Cantor RM, Lange K, Sinsheimer JS (2010) Prioritizing GWAS results: a review of statistical methods and recommendations for their application. Am J Hum Genet 86:6–22
6. Poland JA, Rife TW (2012) Genotyping-by-sequencing for plant breeding and genetics. Plant Genome 5:92–102
7. Pritchard JK, Donnelly P (2001) Case-control studies of association in structured or admixed populations. Theor Popul Biol 60:227–237
8. Marchini J, Cardon LR, Phillips MS, Donnelly P (2004) The effects of human population structure on large genetic association studies. Nat Genet 36:512–517
9. Price AL, Zaitlen NA, Reich D, Patterson N (2010) New approaches to population stratification in genome-wide association studies. Nat Rev Genet 11:459–463
10. Larsson SJ, Lipka AE, Buckler ES (2013) Lessons from Dwarf8 on the strengths and weaknesses of structured association mapping. PLoS Genet 9:e1003246

Chapter 15

De Novo Transcriptome Assembly in Polyploid Species

Juan J. Gutierrez-Gonzalez and David F. Garvin

Abstract

In the absence of a reference genome, the ultimate goal of a de novo transcriptome assembly is to accurately and comprehensively reconstruct the set of messenger RNA transcripts represented in the sample. Non-reference assembly of the transcriptome of polyploid species poses a particular challenge because of the presence of homeologs that are difficult to disentangle at the sequence level. This is especially true for hexaploid oats, which have three highly similar subgenomes, two of which are thought to be nearly identical. Under these circumstances, most software packages and established pipelines encounter difficulties in rendering an accurate transcriptome because they are typically developed, refined, and tested for diploid organisms. We present a protocol for transcriptome assembly in oats that can be extended both to other polyploids and species with highly duplicated genomes.

Key words Oats, Polyploid, De novo transcriptome assembly, RNAseq, Transcript assembly, Homeolog, Homeoallele

1 Introduction

Understanding the assemblage of genes that are expressed in a particular tissue at a certain time—a transcriptome—is of utmost importance for plant researchers because transcriptomes not only provide information on the genes that are active at a certain stage but also on their expression levels. In the event that a reference genome is not available, a de novo transcriptome assembly has to be constructed. Because of the absence of a reference genome sequence that can guide the process, de novo assemblies are more challenging to achieve in an accurate manner. Polyploids pose an extra challenge for de novo transcriptome assembly because of the presence of subgenomes that are frequently very similar. In hexaploid oats this is particularly relevant due to the high homology among the three constituent subgenomes [1, 2]. In a previous study, sequences for homeologous genes coding for the main steps of the vitamin E biosynthesis pathway were identified [1]. The results drew attention to the high similarity among homeologs, with those deriving from two

Sebastian Gasparis (ed.), *Oat: Methods and Protocols*, Methods in Molecular Biology, vol. 1536,
DOI 10.1007/978-1-4939-6682-0_15,

of the three oat subgenomes being nearly identical. Indeed, the nucleotide sequence identity between the two more similar homeologous transcripts for the different steps of the vitamin E pathway ranged from 98.6 to 99.8%, and from 98.0 to 100% at the amino acid level. Moreover, the identities between the two more similar homeologs compared to the more divergent one ranged from 95.5 to 98.2% and from 95.1 to 99.6%, at the nucleotide sequence and amino acid levels, respectively.

This high degree of similarity makes homeologous oat transcripts difficult both to differentiate from each other, and to identify sequencing errors. Two main sources of errors can affect a de novo transcriptome assembly in oats and in other polyploids with highly similar subgenomes. The first is homeolog collapse, which occurs when multiple homeologs are assembled into a single hybrid transcript. The second is SNP shuffling, and occurs when homeolog-specific SNPs are assembled into a homeolog to which it does not belong. This commonly results in a transcript number lower (homeolog collapse) or higher (SNP shuffling) than the three expected transcript isoforms assembled for the same gene, at least for single copy genes that derive from each subgenome. Thus, to increase the chances for a transcriptome with polyploid characteristics to be accurately assembled, it is essential to have (1) high quality starting materials, including high-quality nonfragmented RNA and, subsequently, long high-quality sequencing reads; (2) adequate read coverage; and (3) appropriate software packages and an analysis pipeline that can discriminate homeolog nucleotide differences and sequencing errors. Here, we have developed and tested a pipeline for the de novo assembly of oat transcriptomes. The protocol uses open source software and can be extended to assemble transcriptomes of other polyploids.

2 Materials

The protocol starts with high-quality RNA extracted from the desired tissue, using standard RNA plant extraction protocols (*see* **Note 1**). An amount between 500 and 1000 ng of total RNA is recommended. For a simple transcriptome assembly, where differential gene expression analysis is not part of the study, replicates are not required. Care must be taken when selecting the plant tissue from which to extract RNA, depending on the planned use of the transcriptome (*see* **Note 2**). For instance, one may want to select several different tissues and/or developmental stages to increase the likelihood of capturing a wide range of expressed genes. Once RNA from the desired tissue(s) is extracted, it is usually sent to a genomics facility for further processing. At this facility, sample cDNA libraries will be constructed and subsequently run on a high throughput sequencing platform. For this protocol we assume that library preparation and sequencing are going to take place in such a sequencing facility.

2.1 RNA Extraction

1. TRIzol reagent (Invitrogen, Carlsbad, CA).
2. RNeasy plant columns (Qiagen, Valencia, CA).
3. RNase-free DNase I (New England BioLabs, Ipswich, MA).

2.2 Library Preparation

Library preparation and sequencing are usually carried out in the core sequencing facility and thus, except for some guiding tips, it will not be described in detail here. For a successful assembly with this protocol, the insert size has to be between 500 and 650 bp long, so that a significant proportion of the paired 300 bp long reads have an overlapping fragment (*see* Fig. 1). To size select, first shear sample single stranded RNA, or its double stranded complementary DNA, with a sonicator (Covaris S220 or similar). If RNA was used, synthesize double stranded cDNA from sheared RNA fragments and carry the samples through the remainder of the Illumina® TruSeq® library preparation procedure: adapter ligation and amplification. Use a Caliper® LabChipXT® fractionation system or similar for interval size selection (*see* **Note 3**).

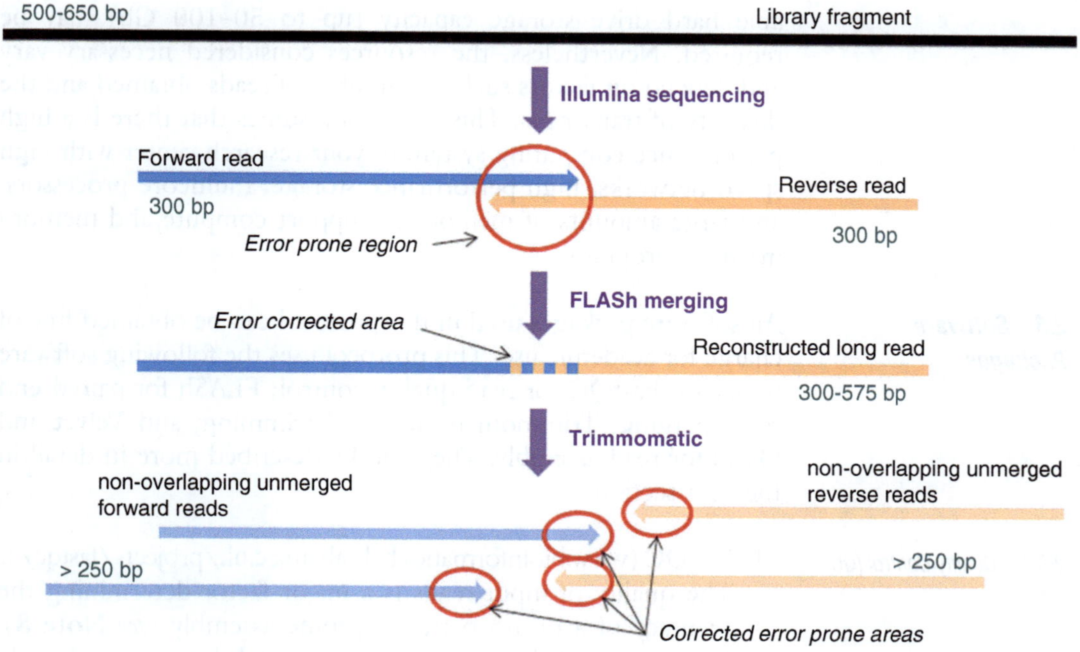

Fig. 1 *Schematic representation of read processing steps.* First, library fragments are sequenced on the Illumina MiSeq instrument to render 300 bp long paired-end reads. Forward reads are in *blue*; reverse in *orange*. The quality within Illumina reads typically decreases toward the end (*red circles*), making these regions error-prone. Second, reads whose 3′-ends overlap are merged into a single longer read by FLASh, which can also correct base errors in the overlapping error-prone region. Third, the remaining unmerged paired-end reads are cleaned of errors and other contaminants with the help of Trimmomatic

2.3 Sequencing

For this protocol we will use Illumina sequencing by synthesis (SBS) chemistry on an Illumina MiSeq® system, which integrates cluster generation, amplification, sequencing, and data analysis into a single instrument (*see* **Note 4**). Libraries have to be sequenced with a 300 bp paired-end run, using V3 chemistry to allow the highest output (up to 15 Gb). Sheared cDNA input can be as little as 10–100 ng.

We believe that for oat transcriptomes, Illumina MiSeq® long paired reads are currently able to yield the best assemblies (*see* **Note 5**). An added value of Illumina reads is that they can also be used later in downstream gene expression quantification analysis. However, sequencing is a rapid-evolving area of research and other promising technologies are on the rise that could also be used in a short future (*see* **Note 6**). This protocol assumes that sequencing is performed in an external genomics facility.

2.4 Computer Resources

This protocol was designed and tested with a 64bit Linux-compatible environment in mind (*see* **Note 7**). Some familiarity with the Unix/Linux command-line user interface would be useful but is not required. A significant amount of memory (200–300 Gb) dedicated over a period extending several days (3–7) and considerable hard drive storage capacity (up to 50–100 Gb) may be required. Nevertheless, the resources considered necessary vary and depend on factors such the number of reads obtained and the diversity of transcripts. This protocol assumes that there is a high performance computing system in your research center with high speed networks, high performance storage, multicore processors, and large amounts of memory to support compute and memory intensive programs.

2.5 Software Packages

All software packages used in this protocol can be obtained free of charge for academic use. This protocol uses the following software packages: FastQC for read quality control; FLASh for paired-end read merging; Trimmomatic for read trimming; and Velvet and Oases for read assembly. They will be described more in detail in the next sections.

2.5.1 Quality Control (QC)

1. FastQC (www.bioinformatics.babraham.ac.uk/projects/fastqc/). The quality of input reads is a major factor determining the accuracy of a de novo transcriptome assembly (*see* **Note 8**). Low quality reads are prone to errors and thus are an obstacle for a precise assembly, especially in polyploid genomes because assemblers might not correctly discriminate homeoallelic SNPs from sequencing errors [3]. Thus, read quality checking should be performed once reads are delivered from the sequencing facility. Overall, Illumina technology produces low-error reads; however, the quality of base calling decreases toward the ends, especially with longer reads. For a 300 bp read, the length used

in this protocol, quality typically decreases greatly in the last 25–50 bases. In diploid organisms, this issue is ameliorated by increased read depth, and thus read errors are corrected to a great extent by assemblers. In polyploid species with very similar subgenomes, base calling accuracy is of paramount importance because assemblers have difficulties in discriminating between subgenome-specific SNPs and read miscalls [3].

2.5.2 Read Elongation

1. FLASh (https://sourceforge.net/projects/flashpage/files/).
   FLASh is a very fast and accurate software tool to merge paired-end reads from next-generation sequencing experiments [4]. FLASh is designed to merge pairs of reads when the original DNA fragments are shorter than twice the length of reads. If a mismatch is present in the overlapping fragment, it can be corrected based on the quality of the base(s) in question. The resulting merged longer reads can significantly improve transcriptome assemblies.

2.5.3 Read Trimming

1. Trimmomatic (http://www.usadellab.org/cms/?page=trimmomatic).
   Trimmomatic is a fast, multithreaded command line quality control tool that can be used to trim and crop Illumina (fastq) poor quality data, as well as to remove adapters and other Illumina-specific sequences [5]. These adapters can pose a significant problem to accurate assemblies as they can be a source for foreign base introgressions.

2.5.4 Assembly of Reads

1. Velvet (http://www.ebi.ac.uk/~zerbino/velvet/).
2. Oases (http://www.ebi.ac.uk/~zerbino/oases/).
   Velvet is a de novo genomic assembler specially designed for short read sequencing technologies [6]. Oases [7] was released as an extension of Velvet to address the particular issues affecting transcriptome assemblies, including (1) transcript sequencing depth potentially varying several orders of magnitude due to differences in gene expression, and (2) the presence of transcript isoforms due to alternative splicing. The Velvet/Oases package is capable of using a combination of long, short, single, and paired-end reads to render high-quality transcriptome assemblies (*see* **Note 9**). For this protocol, Velvet/Oases has to be compiled with specific non-default instructions (*see* **Note 10**).

3 Methods

3.1 RNA Extraction and Library Preparation

1. Grind selected plant tissue using mortar and pestle in the presence of liquid nitrogen.
2. Extract RNA following well-established protocols. We recommend using the TRIzol method, followed by a purification

step with RNeasy plant columns, and digestion with RNase-free DNase I to eliminate traces of DNA. More details can be found in [3].

3. Send high-quality RNA to a core facility for library construction and sequencing. For this protocol we assume that the sequencing facility delivers the resultant sequencing data in two files containing paired end reads: `MiSeq_R1_pair.fastq` and `MiSeq_R2_pair.fastq`, for forward and reverse pairs, respectively.

3.2 Read QC

1. Create a working directory (`MiSeq_reads`) where reads will be copied into.

```
% mkdir MiSeq_reads
% cd MiSeq_reads
```

2. Download and install FastQC (www.bioinformatics.babraham.ac.uk/projects/fastqc/).
3. Run it on the file reads.

```
% /path-to-fastqc-directory/fastqc MiSeq_R1_
pair.fastq MiSeq_R2_pair.fastq
```

The symbol % represents the command line prompt. Substitute `/path-to-fastqc-directory/` with the absolute path to where `fastqc` has been installed. The QC step will provide some basic statistics and other important information. For instance, this will permit an assessment of the presence of sequence contaminants, such as Illumina adapters. Reads should come free of such contaminants from the sequencing facility. Without doubt, apart from the presence of contaminants, the most important piece of information to obtain after running `fastqc` is the per base sequence quality, which will give us a graphic representation of the average base quality along the read and the point at which quality drops below a value that will compromise accurate assembly. This information will be used to modify parameters in downstream software.

3.3 Trimmomatic (Optional)

Should contaminants be detected in the reads, they will have to be removed before the next step, using specialized software (*see* Subheading 3.5). Reads should also go through this optional step if their average quality is low. Because read quality is a key component for an accurate assembly, it is not advisable to proceed with the protocol if read quality is too low. In this case, rerunning the sequencing step is generally recommended.

3.4 Merging Paired Reads with FLASh

1. Download and install FLASh (https://sourceforge.net/projects/flashpage/files/).
2. Run the following command:

```
% /path-to-flash-directory/flash \
```

```
MiSeq_R1_pair.fastq MiSeq_R2_pair.fastq \
-m 25 -M 250 -x 0.1 -o flash.out 2>&1 | tee flash.
   log
```

The first two files are the forward and reverse read input files. The backslash "\" character means that we are continuing the command on the next line, and it is used in scripting languages to improve readability. "m" (min-overlap) is the minimum required overlap length between two reads. Adjust this parameter according to the average insert size and read quality toward the end. Lower values increase the number of extended reads but at the risk of also increasing incorrectly merged pairs. We do not recommend m to be set below 15–20. "M" (max-overlap) is the maximum overlap length expected in 90% read pairs. "x" (max-mismatch-density) is the maximum allowed ratio between the number of mismatched base pairs and the overlap length. Adjust this parameter according to the average read quality toward the ends of the reads. Increasing x increases the number of correctly merged read pairs, but at the expense of increasing the number of incorrectly merged pairs. The string after "o" (flash.out) indicates the prefix for the output files. The "2>&1" operator redirects standard error to the standard output stream (screen). Here, it is piped with "|" to divert a copy of the standard output stream to the intermediate file flash.log.

FLASh will produce three main output files with reads, along with several statistic and error files: flash.out.extendedFrags.fastq: file containing the merged reads; flash.out.notCombined_1.fastq: file with forward reads of the pairs that were not merged; and flash.out.notCombined_2.fastq: file with reverse reads of the pairs that were not merged.

3.5 Trimmomatic

1. Download and install Trimmomatic (http://www.usadellab.org/cms/?page=trimmomatic).
2. Run the following command on the non-merged Illumina paired reads:

```
% java -jar $TRIMMOMATIC/trimmomatic.jar PE
   -threads 4 -phred33 \
-trimlog trimmomatic.log \
flash.out.notCombined_1.fastq   flash.out.not-
   Combined_2.fastq \
flash.out.notCombined_1.PE.fastq \
flash.out.notCombined_1.SE.fastq \
flash.out.notCombined_2.PE.fastq \
flash.out.notCombined_2.SE.fastq \
ILLUMINACLIP:$TRIMMOMATIC/adapters/TruSeq2-
   PE.fa:2:30:10:2:true \
```

```
LEADING:10 TRAILING:10 SLIDINGWINDOW:4:15
   MINLEN:250
```

This will do the following:

- Remove adapters and other Illumina-specific sequences (`ILLUMINACLIP:$TRIMMOMATIC/adapters/TruSeq2-PE.fa:2:30:10:2:true`). This command will clip reads according to the adapters and other sequences in `TruSeq2-PE.fa`, maximum mismatch count (`2`), palindrome clip threshold (`30`), simple clip threshold (`10`), minimum adapter length (`2`) to allow shorter fragments to be removed, and keep both reads (`true`) in the cases in which forward and reverse contain the same sequence information, albeit in reverse complement. Most likely you will not have to modify this part.
- Remove leading low quality or N bases (below quality 10) (`LEADING:10`).
- Remove trailing low quality or N bases (below quality 10) (`TRAILING:10`).
- Scan the read with a 4-base wide sliding window, cutting when the average quality per base drops below 15 (`SLIDINGWINDOW:4:15`).
- Eliminate reads below the 250 bases long (`MINLEN:250`).

You may want to modify the numbers in `LEADING`, `TRAILING`, and `SLIDINGWINDOW` depending on the quality of your reads. Additionally, you may want to adjust `MINLEN` by using a higher or lower threshold if the quality of reads is high or low, respectively.

Two input files with the forward (`flash.out.notCombined_1.fastq`) and reverse (`flash.out.notCombined_2.fastq`) reads to clean are needed. They contain the reads not merged by FLASh in the previous step. Trimmomatic will produce five output files: `trimmomatic.log` is the log file, which includes information such as the read name, the surviving sequence length, the number of bases trimmed from the start, and the number of bases trimmed from the end; `flash.out.notCombined_1.PE.fastq` and `flash.out.notCombined_2.PE.fastq` are the forward and reverse paired read files that survived the trimming process as pairs. Finally, `flash.out.notCombined_1.SE.fastq` and `flash.out.notCombined_2.SE.fastq` are the unpaired read files whose matching pair did not pass the trimming process, and thus are single read files now.

3.6 Read Assembly

1. Download and install Velvet (https://www.ebi.ac.uk/~zerbino/velvet/) and Oases (http://www.ebi.ac.uk/~zerbino/oases/). The Velvet/Oases package is comprised of three independent modules that have to be run in a sequential order: *velveth*, *velvetg*, and *oases*. Because of the large size of the intermediate files created in the process, we strongly

recommend testing Velvet/Oases in a scratch directory and copy the informative files (typically just `transcripts.fa`) to a more permanent directory when you are satisfied with the outcome. For large datasets some recommendations can be followed to speed up the process (*see* **Note 11**).

2. Run *velveth* with the following command:

```
cd scratch_directory
% /path-to-velvet-directory/velveth Miseq_
   assembly 149 \
-long -fastq /MiSeq_reads/flash.out.extended-
   Frags.fastq \
-shortPaired -separate -fastq /MiSeq_reads/
   flash.out.notCombined_1.PE.fastq \ /MiSeq_
   reads/flash.out.notCombined_2.PE.fastq \
-short -fastq /MiSeq_reads/flash.out.notCom-
   bined_1.SE.fastq \
/MiSeq_reads/flash.out.notCombined_2.SE.fastq
```

Where `Miseq_assembly` is the name of the folder we want Velvet/Oases to store the results. Probably, the most important parameter to specify is the *k-mer* value (`149` in this case). We recommend higher (**125–149**) than typical values suggested for diploid species (*see* **Note 9**). We also recommend testing several *k-mers*. The next entries are self explanatory. They describe the type of reads (`-long`, `-short`, `-shortPaired`) and the files that contain them (for more detailed parameter description check Velvet manual, http://www.ebi.ac.uk/~zerbino/velvet/Manual.pdf).

3. Run *velvetg* with the following command:

```
% /path-to-velvet-directory/velvetg Miseq_
   assembly \
-read_trkg yes -min_pair_count 2 -cov_cutoff
   auto -ins_length 650 \
-exp_cov 5 -max_gap_count 1 -min_contig_lgth
   300 -conserveLong yes
```

Change the `ins_length` according to your paired-end library average insert size (*see* **Note 3**). Also, modify the expected *k-mer* coverage (`exp_cov`) accordingly. The rest of the parameters are probably best left untouched (for more detailed parameter description check Velvet manual).

4. Run *oases* with the following command:

```
% /path-to-oases-directory/oases Miseq_assem-
   bly \
-min_trans_lgth 300
```

where `min_trans_lgth` is the desired minimum transcript length. After a successful run (*see* **Note 12**) several output files are created, of which `transcripts.fa` is the most useful for use as it contains the assembled homeologs. `transcripts.fa` is a `fasta` file with transcript isoforms organized in *Loci*, where individual transcript isoforms within a *Locus* might represent homeologs but also splice variants. The `fasta` header for each transcript appears as:

```
>Locus_x_Transcript_y/Y_Confidence_z_Length_L
```

where "x" is the Locus unique identification number; "y" is the transcript isoform number among the "Y" isoforms in this `Locus`; "z" is a number between $0 \leq z \leq 1$ (ideally we want z to be as close as 1 as possible); and "L" is the length of that particular transcript isoform. For oats, in the ideal situation where three putative homeologs are assembled, we should see something like this:

```
>Locus_x_Transcript_1/3_Confidence_z_Length_L
>Locus_x_Transcript_2/3_Confidence_z_Length_L
>Locus_x_Transcript_3/3_Confidence_z_Length_L
```

Of course, the values of `z` and `L` could be different among the three isoforms (homeologs).

4 Notes

1. RNA will be extracted using standard plant RNA extraction protocols. RNA purity and integrity is crucial for a successful read synthesis. We recommend using the TRIzol® method (Invitrogen, Carlsbad, CA), followed by purification with RNeasy plant columns (Qiagen, Valencia, CA). Digestion with RNase-free DNase I (New England BioLabs, Ipswich, MA) to eliminate DNA traces is highly recommended. Prior to library construction, determine RNA quality and integrity with RNA6000 Nano Assay on the Agilent 2100 Bioanalyzer™ (Agilent Technologies Inc, Santa Clara, CA). A RIN value above 7 is desirable.
2. In order to increase transcript diversity, samples from several biological replicates and/or developmental stages may be collected and pooled. If the ultimate purpose for the transcriptome is to be used as a foundation for expression analysis, biological replicates are a must. Long paired-end reads are also highly recommended, and used in this protocol, because they are more likely to map to a single homeolog. If the final purpose of the transcriptome is expression studies, very strict or even zero-tolerance parameters are desired when aligning reads to the assembled transcriptome for read counting.
3. The more stable cDNA is preferred over unstable, single stranded RNA as starting material for shearing. The desired target fragment size of 500–650 bp reflects the length after subtracting off the Illumina adapter/barcode sequence of 120 bp. Thus, the fragment to select in the fractionation system must be in the range of 620–770. For a correct shearing and library representation, the starting RNA or cDNA material should be at least 500 ng; however, because of the subsequent

size selection step an amount of 10 μg or higher is recommended (also *see* **Note 8**).

4. Illumina sequencing by synthesis (SBS) chemistry is the most widely adopted next-generation sequencing technology because of the low sequencing errors and high-throughput. Illumina MiSeq® system using V3 chemistry is capable of generating 20–25 M reads on average and more than 12 Gb (up to 15 Gb) of data for each 2 × 300 base pair run. The MiSeq also has the ability to multiplex several samples per run. Currently, Illumina sequencing delivers the highest yield of error-free data. The typical error rate for an entire Illumina Miseq 300 bp long read may vary between 0.5 and 2 % and depends on library type, run quality, and a host of other factors.
5. In our experience, combining different technologies, such as Roche-454 and Illumina, to perform what is known as a *hybrid assembly* is not recommended for polyploid species, where the similarity among subgenomes can confound assemblers. This is because different technologies have distinct error patterns. Assembler algorithms are typically designed with a particular type of sequencing technology in mind and to deal with specific error patterns.
6. Single molecule real-time (SMRT®) sequencing (PACBIO®, Pacific Biosciences of California, Inc) is a promising technique that can develop reads as long as 10 kb on average. However, its single pass accuracy of about 85 % is probably still too high to discriminate oat homeolog single nucleotide variants from sequencing errors. Should SMRT sequencing improve accuracy in the next years, it will represent an interesting alternative to Illumina technology for transcriptome assemblies and it will be worth considering.
7. Although the protocol might work on other computer systems, particularly on a 32-bit machine, no other system has been tested by us. Running a 32-bit machine might find memory to be a limiting factor.
8. Two key factors affecting the quality of an assembly are the quality of sequence reads and the level of read duplication. Read errors can confound assemblers and are especially problematic in polyploid organisms. Read duplication is highly dependent on the quality of the input cDNA and library preparation. To minimize these errors and increase efficiency of sequencing and rare gene discovery, we recommend (1) conducting library preparation from enough RNA/cDNA starting material and perform some type of library normalization so there is no over-representation of certain abundant transcripts; and (2) avoid a large variance in the fragment size selected such that smaller fragments become over-represented. Normalization

methodologies are desirable in situations where the actual RNA counts are unimportant (de novo transcriptome), but not appropriate when transcript quantification is an objective (RNAseq).

9. There are two major classes of assembling algorithms: the *de Bruijn graphs* and the overlap-layout-consensus [8]. Velvet/Oases is an example of the first type. For assemblers that use *de Bruijn graphs* methods, the quality of a de novo assembly is highly dependent on the selected *k-mer* value. The optimum value depends on the sequencing depth, the read error rate, and the complexity of the genome/transcriptome to be assembled [9]. For transcriptomes, where coverage is not uniform, the dilemma is choosing between a higher *k-mer* length that will result in a more contiguous assembly of highly expressed transcripts; or a lower *k-mer*, which theoretically will favor weakly expressed transcripts but will lead to the assembly of numerous and highly fragmented transcripts [7]. For polyploids, we do not recommend performing a ***merged assembly***, that is, running velvet with different values of *k* and merge the results of all assemblies into a single nonredundant assembly. According to our benchmarks, merging tends to produce highly fragmented and mixed assemblies. Neither do we recommend the use of overlap-layout-consensus assemblers for transcriptomes in polyploid species, as they have the tendency to mix fragments from different homeologs.

10. By default, Velvet is compiled with `MAXKMERLENTH` of 31 bp. Thus, hash-lengths (*k-mers*) are limited to 31 bp. For the Velvet/Oases package to be able to run with the parameter values specified in this protocol, it has to be compiled with the following instructions, so that both long *k-mers* and long sequences can be used:

```
make 'CATEGORIES=7' 'MAXKMERLENGTH=149' 'LONGSEQUENCES=1'
    'BIGASSEMBLY=1'
```

11. To speed up the assembly process a couple of approaches may be taken. First, you may use multithreading, which allows a program to make use of multiple CPU cores on the same machine. To turn on multithreading in Velvet use the option make `'OPENMP=1'` at compilation. OpenMP allows a program to make use of multiple CPU cores on the same machine. You might have to set the environment variables `'OMP NUM THREADS'` and `'OMP THREAD LIMIT'`. See the Velvet documentation for more information. Second, you can significantly speed up Velvet by working with a binary version of sequences. For this, simply add the `-create_binary` parameter to the *velveth* command:

```
% /path-to-velvet-directory/velveth Miseq_
  assembly 149 -create_binary (...)
```

12. There is not a standard reliable method to validate the results of a de novo transcriptome assembly. Nevertheless, at least three approaches can be taken to assess the quality of a transcriptome without the need for a reference genome. First, software packages such as TransRate [10] can be used for reference-free quality assessment of de novo transcriptome assemblies. Second, quality can be assessed by aligning the de novo transcriptome to a well-curated set of reference ESTs present in public databases (GenBank, EMBL, or DDBJ). The last but the least precise method is by aligning the transcriptome to a well-conserved set of genes from close relatives.

Acknowledgments

Mention of trade names or commercial products in this article is solely for the purpose of providing specific information and does not imply recommendation or endorsement by the U.S. Department of Agriculture. This work was carried out in part using computing resources at the University of Minnesota Supercomputing Institute. The authors thank the Biomedical Genomics Center, University of Minnesota for the testing of different protocols for library preparation and sequencing reactions.

References

1. Gutierrez-Gonzalez JJ, Garvin DF (2016) Subgenome-specific assembly of vitamin E biosynthesis genes and expression patterns during seed development provide insight into the evolution of oat genome. Plant Biotechnol J. doi:10.1111/pbi.12571
2. Gutierrez-Gonzalez JJ, Garvin DF (2011) Reference genome-directed resolution of homologous and homeologous relationships within and between different oat linkage maps. Plant Genome 4:178–190
3. Gutierrez-Gonzalez JJ, Zheng JT, Garvin DF (2013) Analysis and annotation of the hexaploid oat seed transcriptome. BMC Genomics 14:471
4. Magoc T, Salzberg S (2011) FLASh: fast length adjustment of short reads to improve genome assemblies. Bioinformatics 27(21):2957–2963
5. Bolger AM, Lohse M, Usadel B (2014) Trimmomatic: a flexible trimmer for Illumina sequence data. Bioinformatics 30(15):2114–2120
6. Zerbino DR, Birney E (2008) Velvet: algorithms for the novo short read assembly using de Bruijn graphs. Genome Res 18:821–829
7. Schulz MH, Zerbino DR, Vingron M, Birney E (2012) Oases: robust *de novo* RNAseq assembly across the dynamic range of expression levels. Bioinformatics 28:1086–1092
8. Li Z, Chen Y, Mu D, Yuan J, Shi Y, Zhang H, Gan J, Li N, Hu X, Liu B, Yang B, Fan W (2012) Comparison of the two major classes of assembly algorithms: overlap-layout-consensus and de-Bruijn-graph. Brief Funct Genomics 11(1):25–37
9. Surget-Groba Y, Montoya-Burgos JI (2010) Optimization of de novo transcriptome assembly from next-generation sequencing data. Genome Res 20:1432–1440
10. Smith-Unna R, Boursnell C, Patro R, Hibberd JM, Kelly S (2015) TransRate: reference free quality assessment of *de-novo* transcriptome assemblies. Genome Res 26(8):1134–1144

Part V

Proteomics Methods

Chapter 16

Isolation of Oat (*Avena sativa* L.) Total Proteins and Their Prolamin Fractions for 2D Electrophoresis

Dorota Nałęcz, Marta Dziuba, and Iwona Szerszunowicz

Abstract

Appropriate sample preparation is essential to obtaining good results of two-dimensional gel electrophoresis (2-DE). For various reasons (particularly phenolic compounds, proteolytic enzymes, and cell-wall mucilages) the extraction of proteins from plant material, among them oat proteins, is difficult. During isolation all soluble substances that may interfere with the analysis (especially isoelectric focusing) are removed, and proteins of interest are separated from the remains. However, the applied procedure of isolation cannot be too extensive, because additional stages cause loss of the proteins.

In this chapter, we describe a simple procedure for the isolation of oat total proteins and their prolamin fractions prior to 2-DE, without necessity of considerable purification. It can be used for oat protein fractionation, measurement of oat protein concentration, and their 2-DE analysis, with particular reference to prolamin fractions. The presented routine includes modified methods of plant seed proteins extraction and sequential Osborne extraction, based on oat protein solubility differences.

Key words Oat proteins, Avenins, Sample preparation for 2-DE, Isolation, Ethanol extraction of prolamins, Determination of protein concentration

1 Introduction

The choice of a proper protein isolation procedure (including sample preparation and selection of the one among many different protein extraction methods) is the key to a successful laboratory experiment [1], especially with such a sensitive and selective analytical process as 2-DE. Effective preparation of a protein sample for 2-DE requires precise removal of other present compounds prior to electrophoresis (such as nucleic acids, lipids, polysaccharides, phenols, and salts) which bind to proteins and may interfere with 2-DE, and separation of proteins into fractions and single subunits from complex mixtures [2]. The method of preparing a protein sample depends on the type of biological material used as a protein source and characteristic features of proteins, such as solubility, hydrophobicity, molecular weight and size, isoelectric point

Sebastian Gasparis (ed.), *Oat: Methods and Protocols*, Methods in Molecular Biology, vol. 1536,
DOI 10.1007/978-1-4939-6682-0_16,

Table 1
Oat protein fractions according to Osborne protein solubility[a]

Name of protein fraction	% of total protein	MW (kDa)	p*I*	Solubility
Globulins	50–80	20–35 50–60	5.5 and 8–10	Soluble in dilute aqueous salt solutions
Albumins (mostly enzymes)	1–12	14–17 20–27 36–47	4–7	Water soluble
Prolamins (avenins)	4–15	17–35	5–9	Soluble in aqueous ethanol
Glutelins	<10	9	–	Soluble in dilute acids or alkalis

[a]From ref. 6

and charge etc. [3, 4]. Some of protein extraction methods are performed with the use of simple solubilization solutions, and other require buffers with complex mixtures of chaotropic agents, detergents, reducing agents, and ampholytes for protein solubilization (especially from plant tissues) during isolation, and to maintain solubility during the focusing [5].

This chapter describes the procedure for isolation of oat total proteins and their prolamin fractions prior to 2-DE. The presented procedure is a modification of extraction methods described early by Görg et al. [6] with reference to plant seeds (extraction of total proteins and sequential Osborne extraction, based on protein solubility differences) (*see* Table 1) [7]. A properly performed isolation that takes into account all the advice and major sources of problems allows to prepare samples with high concentrated proteins of interest, and to achieve efficient and reproducible 2-DE separation (*see* Chapter 17).

2 Materials

2.1 Equipment

1. Laboratory mill with speed of up to 20,000 × *g* and grinding results of 250-μm particle size (*see* **Note 1**).
2. 250-μm mesh sieve.
3. Solvent vacuum microfiltration assembly.
4. Centrifuge with a cooling system, capable of at least 40,000 × *g*. The centrifuge tube size depends on the centrifuge and rotor. For extractions of oat total proteins we use tubes made of polycarbonate 10 mL, incl. a screw cap, 16.1 i.d. (inside diameter) × 81.1 mm, and for extraction of oat prolamins we use tubes made of polycarbonate 30 mL, incl. a screw cap, 25.3 i.d. (inside diameter) × 92 mm (*see* **Note 2**).

5. Needle sample concentrator (Techne, Stone, UK) (*see* **Note 3**).
6. Freeze Dry System (Labconco, Kansas City, USA) (*see* **Note 4**).
7. VIS spectrophotometer (set to 595 nm) and polystyrene cuvettes with 1 cm path length matched to a laboratory spectrophotometer (*see* **Note 5**).
8. Grade 1 Whatman filter (or equivalent) and funnel for dye reagent preparation (*see* **Note 6**).

2.2 Cereal Material

Oat (*Avena sativa* L.) seeds from any cultivar are needed.

2.3 Reagents

We recommend filtration of buffer solutions after their preparation with the use of a solvent vacuum microfiltration assembly to remove solid micro impurities. Important: Preparation of all solutions should be done very carefully because of many hazardous (poisonous and corrosive) ingredients, for example: Tris(hydroxymethyl) aminomethane (other names: Tris, Tris buffer, Tris base, Trizma, Trisamine), urea, dithiothreitol (DTT), phenylmethylsulfonyl fluoride (PMSF), 3-[(3-cholamidopropyl)-dimethylammonio]-1-propanesulfonate (CHAPS), hydrochloric acid (HCl). Wear an appropriate laboratory coat, gloves, and safety glasses.

2.3.1 Extraction of Oat Total Proteins (See Note 7)

1. Extraction buffer A: 6 M urea, 1 % (w/v) CHAPS, 0.5 % (w/v) DTT, 0.5 % (v/v) immobilized pH gradient (IPG) buffer pH 3–10, 1 % (w/v) polyvinylpolypyrrolidone (PVPP), 1 mM PMSF. To prepare 50 mL of extraction buffer A, dissolve 60 g of urea in water (*see* **Note 8**) and adjust the volume to 50 mL, stir, and filter. Add 0.5 g of CHAPS, 0.25 g of DTT, 0.25 mL of IPG, 0.5 g of PVPP (*see* **Note 9**), and 8.7 mg of PMSF (*see* **Note 10**) to ~48 mL of the filtered urea solution. Stir and adjust the volume to 50 mL. The buffer must be freshly prepared and used immediately or stored in 1 mL aliquots at −80 °C for up to several months.
2. Anhydrous propan-2-ol or ethanol.

2.3.2 Extraction of Oat Prolamins (See Note 7)

1. Extraction buffer B: 50 mM Tris–HCl buffer pH 8.8, 1 % (w/v) PVPP, 1 mM PMSF. To prepare 50 mL of extraction buffer B, dissolve 0.3 g of Tris base in ~40 mL of water. Adjust pH to 8.8 with HCl (*see* **Note 11**), filter, add 0.5 g of PVPP, 8.7 mg of PMSF, stir and adjust the volume to 50 mL with water. Store the buffer at 4 °C.
2. 75 % (v/v) ethanol.

2.3.3 Determination of Protein Concentration [8]

1. Bio-Rad protein assay kit with lyophilized bovine serum albumin standard (Bio-Rad, California, USA). Store the dye reagent concentrate and lyophilized bovine serum albumin standard at 4 °C. If the standard is not used within 60 days, it

should be aliquoted and frozen at −20 °C for maximum 5 years. The dye reagent concentrate is stable for 1 year when kept at 4 °C. After dilution the reagent may be used for approximately 2 weeks.

2. Rehydration buffer stock solution: 7 M urea, 2 M thiourea, 4 % (w/v) CHAPS. To prepare 25 mL of rehydration stock solution, dissolve 10.5 g of urea, 3.8 g of thiourea, and 1 g of CHAPS in water, stir, adjust the volume to 25 mL, and filter. Divide all volume on 3 mL aliquots, and store at −20 °C (*see* Chapter 17).

3 Methods

3.1 Preparation of Oat Flour Sample

1. Grind the necessary amount of oat seeds (*see* **Note 12**) in a laboratory mill.
2. Sieve flour through a 250-μm mesh sieve.
3. After milling store the flour in a refrigerator (2–8 °C) until it is used for extraction.

3.2 Extraction of Oat Total Proteins

1. Add 5 mL of extraction buffer A to 500 mg of oat flour and carefully mix with the use of a spatula, and vortex.
2. Perform extraction for 1 h at room temperature, and vortex every 15 min.
3. After this time, centrifuge the extract (40,000 × *g*, for 1 h, at 15 °C).
4. Remove the obtained supernatant (containing oat proteins) with a pipette and store it at −80 °C.
5. Lyophilize frozen supernatants at −54 °C for 12 h and store lyophilizates at −80 °C.
6. Measure the protein concentration in a sample solution as described in Subheading 3.4.
7. Subject oat protein extracts (**step 4**) or oat protein lyophilizates (**step 5**) to 2-DE (*see* Chapter 17) (*see* **Notes 13** and **14**).

3.3 Extraction of Oat Prolamins

1. Add 16 mL of extraction buffer B to 4 g of oat flour and carefully mix with the use of a spatula, and vortex.
2. Perform extraction for 1 h at 4 °C with vortexing every 15 min.
3. After this time, centrifuge the extract (20,000 × *g*, for 20 min, at 15 °C).
4. Remove supernatant containing albumin and globulin fractions. You can lyophilize it and store at −80 °C.
5. Repeat subsequently the procedure of extraction and centrifugation one more time.

6. Wash the pellet with 20 mL of deionized water, by mixing for 2–3 min with a spatula, and vortex.
7. Centrifuge (under conditions as specified above).
8. Extract prolamins twice, by pouring 16 mL of 75 % (v/v) ethanol over the pellet, and vortexing.
9. Stir the samples on the lab shaker for 2 h at room temperature with vortexing every 30 min.
10. After centrifugation (under conditions as specified above), vaporize the obtained supernatant containing oat prolamins in a needle concentrator to 1/4 of the initial volume of the extract.
11. Store evaporated supernatants (containing oat prolamins) at −80 °C.
12. Lyophilize frozen supernatants at −54 °C for 12 h and store lyophilizates at −80 °C.
13. Measure the protein concentration in a sample solution as described in Subheading 3.4.
14. Subject oat protein extracts (**step 10**) or oat protein lyophilizates (**step 12**) to 2-DE (*see* Chapter 17) (*see* **Notes 13** and **14**).

3.4 Protein Concentration Measurement

For measuring the concentration of total proteins and prolamins, a standard procedure based on the Bradford dye-binding method [8] according to Instruction Manual, Bio-Rad Protein Assay is used. This is a simple colorimetric assay based on the color change of Coomassie Brilliant Blue (CBB) G-250 dye in response to various concentrations of protein (*see* **Note 15**). It involves the addition of an acidic dye to a protein solution, and subsequent absorbance measurements at 595 nm with a spectrophotometer. Comparison to an established standard curve provides a relative measurement of protein concentration.

1. In order to measure protein concentration in your oat protein sample prepare five dilutions of a protein standard solution (BSA, ~1.5 mg/mL, including in a protein assay kit) in the range from 0.2 to 0.9 mg/mL.
2. Pipette 20 μL of each standard and sample solution (e.g., 0.4 mg of lyophilizate/mL of rehydration buffer stock solution, do not dilute liquid extract solution) to a clean Eppendorf tube and add 1 mL of a diluted dye reagent to each tube.
3. Be aware of sample amount dependence on the protein concentration in it and on the used protein assay kit. If it turns out that protein concentration in the analyzed oat sample is too high or too low (e.g., for grain subject to thermal shock) and the absorbance of the obtained solution is outside the calibration curve, then you need to take an appropriately smaller or bigger amount of the sample (dilute or enrich the sample).

4. Incubate prepared samples at room temperature for 10 min, and measure absorbance at 595 nm.

4 Notes

1. For uniform extraction, it is important to obtain particles of similar size. Particles that do not pass through the sieve should be transferred back into the mill and ground again. We recommend laboratory mill type IKA A 10 (Labortechnik, Staufen, Germany).
2. We recommend centrifuge type Sigma 3K30 (Sigma Laboratory Centrifuges, Osterode am Harz, Germany).
3. A needle sample concentrator requires additionally a simple gas delivery system for solvent evaporation. The producer recommends the use of inert gases with the sample concentrator, with pure nitrogen being the most commonly used. Reactive gases such as oxygen should not be used, as they can react with the solvent or chemical to produce hazardous or inflammable products. The concentrator passes an inert gas over the surface of multiple samples in a block heater via stainless steel needles at once. Optionally, a centrifugal evaporator for the efficient and gentle evaporation of solvents from many samples can be used. Important: After ethanol's extraction of oat prolamins you must remove the ethanol from the extract prior to freeze drying. Highly concentrated alcohol solutions do not freeze at −54 °C (a freezing point of pure ethanol is −114 °C), so it must be removed for proper sublimation during lyophilization.
4. A Freeze Dry System, its size and design will affect the freeze drying speed (time) and freeze drying capacity (the volume of a sample that can be freeze dried at one time). It is dependent on the volume, surface area and thickness of the sample, condenser temperature, and vacuum obtained. The greater the surface of the sample, the greater the rate of freeze drying and for a given surface area the thicker the sample, the slower the rate of freeze drying.

   The samples should be frozen to maximize the surface area and minimize the thickness of the sample. The volume of the used freeze dry flask (or glass bottle, beaker, ampule, vial) should be two to three times the volume of the sample. Never fill a flask to more than 1/3 of its total capacity. To allow air circulation cover a flask with muslin or intralining. Be cautious, glass flasks can break during the freeze drying process. Hairline cracks or nicks in the glassware invisible to the naked eye can cause cracking under cold temperatures, or when subject to vacuum. Important: samples should be prefrozen in the freezer (at −80 °C) before placing them in the freeze dryer for more efficient freeze drying.

5. You should use disposable polystyrene cuvettes, good in the visible region, which absorb less blue dye than glass cuvettes. Make sure that cuvettes are suitable for absorbance measurements, and that the cuvette surface is kept clean on the outside. Remember that adding cold solutions into the cuvette can cause fogging of the cuvette surface, and warming of a solution in the cuvette can cause air bubbles.
6. Prepare to the analysis Coomassie Brilliant Blue (CBB) G-250 dye reagent by dilution with H_2O and filter it through a Whatman filter, grade 1 (or equivalent) to remove particles.
7. During or after cell lysis, interfering compounds such as proteolytic enzymes, salts, lipids, polysaccharides, and plant phenols must be inactivated or removed. We perform extraction of oat proteins and their prolamin fractions based on Simpson's Laboratory Manual [6] with the use of modified urea lysis buffer (extraction buffer A) and Tris–HCl buffer, pH 8.8 (extraction buffer B), respectively. To the buffers we add 1% (w/v) PVPP to remove phenols present in oat material. Oat seeds contain simple phenolic acids, flavonoids, as well as polyphenols specific for oat called avenanthramides [9, 10]. They may interact with proteins, strongly interfere with IEF during their 2-DE separation, and give rise to horizontal streaks in 2-DE gel patterns. To protect isolated oat proteins from the action of proteases and other protein-modifying enzymes, you should use a protease inhibitor that leads to their inactivation, for example PMSF, an irreversible inhibitor that inactivates serine and some cysteine proteases. Otherwise, it may result in an artifactual spot in 2-DE gels [2, 6]. We use PMSF despite some of their disadvantages as limited solubility (*see* **Note 10**), instability in aqueous solutions and toxicity, partly because of its relatively low price. You can use more expensive but nontoxic, water-soluble, and water-stable PMSF alternatives such as Pefabloc® SC protease inhibitor (Pentapharm, Basel, Switzerland).

   Additional substances such as detergents, added to improve one aspect of the workflow (for example Triton X-100 is useful for the solubilization of aggregating proteins), often interfere with later steps (for example Triton X-100 may interfere with 2D Quant Assay during measurement of oat protein concentration).
8. To avoid modification of proteins, never heat a sample after adding urea. Elevated temperatures cause urea degradation to isocyanate, which modifies proteins by carbamylation. When the sample contains urea, it must not be heated above 37 °C.
9. PVPP is insoluble in water, though it still absorbs water and swells very rapidly. You can add solid insoluble PVPP to the extraction buffer before use; allow to hydrate. Store PVPP at room temperature.

10. Due to little water solubility of PMSF we recommend according to Sigma-Aldrich (Darmastadt, Germany) the preparation of a stock solution: 100 mM in anhydrous propan-2-ol or anhydrous (99.8 %, not 96 %) ethanol and the use of the final concentration of 0.1 mM in the lysis buffer. The organic solvent at this concentration has no effect on the proteins. Add PMSF solution to cold lysis buffer right before use because it is unstable, especially at high pH. The stock can be stored up to 9 months in aliquots at −20 °C.

11. We adjust pH of an extraction buffer B to 8.8 by using concentrated (11.5 M) HCl at first to partially narrow the gap from the starting pH to the required pH. Then, it would be better to use a series of HCl (e.g., 6 and 1 M) with lower ionic strengths to avoid a sudden drop in pH below the required pH.

    Put the beaker with the buffer solution on a magnetic stirrer, and keep it during the adjusting of pH. Remember that the pH of Tris solution changes with temperature. Leave it at room temperature for about 30 min. Wait for the pH meter reading to be stabilized.

12. The necessary amount of oat seeds is dependent on the purpose of performed extraction (measurement of protein concentration, protein fractionation, 2-DE analysis) and the final protein concentration in the obtained extract. If the concentration of protein is small, more extract/lyophilizate will be needed for 2-DE. Because of low content of prolamins in the oat seeds you should weigh out more flour for extraction than for other cereals.

13. Oat protein extracts can be used for 2-DE immediately after protein extraction or after their previous lyophilization.

14. Additional steps of the isolation procedure can improve the quality of the final experimental result, but can also be the possible causes of selective protein loss. When making decisions concerning the method of protein purification, it is very important to apply a procedure that uses the smallest number of techniques, but provides a product with the required degree of purity. This results, among other things, in the risk of losing considerable amounts of protein and its activity at the individual steps of purification, as well as the possibility of substantial extension of time, thus increasing the final costs of the entire purification process [2, 11].

    We recommend avoiding the use of such techniques as solid phase extraction (SPE) for purification of plant proteins. It is an increasingly useful sample preparation technique for sample extraction, concentration, and cleanup (removing particles with small molecular weight, among others salts and phenols). But it is dependent on what protein we are purifying and what source material we use. We used a set for the solid phase extraction with the tubes containing silica-based reversed phase LC-18 packings and with solutions containing acetonitrile, trifluoroacetic acid, and water, directly after the extraction of plant proteins (oat, pea, buckwheat). Analysis of 2-DE protein images showed much fewer protein spots for SPE cleaned proteins in comparison to not-cleaned proteins. This result confirms the loss of the part of proteins during the solid phase extraction.

Application of the solid phase extraction leads to obtaining a well-focused image of main protein fractions (with the highest contents), but cannot be recommended for low abounded, among them oat prolamins, which constitute only about 15 % of total oat proteins. Also we do not recommend any ready to use cleanup kits for preliminary oat protein sample cleanup because of protein loss.

15. Be aware of Bradford assay dependence on the type of protein sample and the type of tissue. Because the sample may contain substances that interfere with the assay, such as a detergent, you must check their compatibility and permissible concentrations in the producer guide.

Use of CHAPS instead of Triton X-100 enables exact absorbance measurement. We find that 1 % (w/v) Triton X-100 is not compatible with the assay and induces incorrect absorbance reading of oat protein solution. It is known that 1 % (w/v) or above Triton X-100 considerably enhances the absorbance of CBB G-250 dye binding when BSA is used as a standard, resulting in the underestimation of tissue proteins [12]. So, to avoid interference with the protein measurement assay, we recommend using a CHAPS detergent instead of Triton X-100 in a rehydration solution or using 2D Quant Kit from GE Healthcare (Fairfield, USA). It is designed for the accurate determination of protein concentration in harsh solutions of samples to be analyzed by the 2-DE technique, and is based on quantitatively precipitating proteins while leaving interfering substances in the solution. It is compatible with all components of the rehydration solution: detergents such as 4 % (w/v) CHAPS, 2 % (w/v) SDS, and 1 % (w/v) Triton X-100, reductants such as 1 % (w/v) DTT, chaotropes (8 M urea, 2 M thiourea), and carrier ampholytes such as 2 % (v/v) IPG buffer.

Acknowledgment

This work was supported by the Ministry of Science and Higher Education, Poland within project No NN 312 286333.

References

1. Namieśnik J, Szefer P (2008) Preparing samples for analysis-the key to analytical success. Ecol Chem Eng S 15(2):167–244
2. GE Healthcare Life Sciences (2015) Protein sample preparation. Handbook. http://www.gelifesciences.com/webapp/wcs/stores/servlet/catalog/en/GELifeSciences-pl/service-and-support/handbooks/. Accessed 9 Aug 2015
3. Pavoković D, Križnik B, Krsnik-Rasol M (2012) Evaluation of protein extraction methods for proteomic analysis of non-model recalcitrant plant tissues. Croat Chem Acta 85(2):177–183
4. Martínez-Maqueda D, Hernández-Ledesma B, Amigo L et al (2013) Extraction/fractionation techniques for proteins and peptides and protein digestion. In: Toldrá F, Nollet LML (eds) Proteomics in foods: principles and applications, vol 2, Food microbiology and food safety. Springer Science+Business Media, New York, pp 21–50
5. Rabilloud T, Luche S, Santoni V et al (2007) Detergents and chaotropes for protein solubilization before two-dimensional electrophoresis. In: Hervé T, Zivy M (eds) Plant proteomics, vol 355, Methods in molecular biology. Humana Press, New York, pp 111–119
6. Görg A, Drews O, Weiss W (2004) Separation of proteins using two-dimensional gel electrophoresis. In: Simpson RJ (ed) Purifying proteins for proteomics. A laboratory manual. Cold Spring Harbor Laboratory Press, New York
7. Klose C, Arendt EK (2012) Proteins in oats; their synthesis and changes during germination: a review. Crit Rev Food Sci Nutr 52:629–639
8. Bradford MM (1976) A rapid and sensitive method for the quantitation of microgram quantities of protein utilizing the principle of protein-dye binding. Anal Biochem 72:248–254
9. Meydani M (2009) Potential health benefits of avenanthramides of oats. Nutr Rev 67:731–735
10. Boz H (2015) Phenolic amides (Avenanthramides) in oats-a review. Czech J Food Sci 33(5):399–404
11. Rosiński M, Piasecka-Kwiatkowska D, Warchalewski J (2005) Przegląd metod separacji i oczyszczania białek przydatnych w badaniach i analizie żywności (The review of protein separation and purification methods appearing useful in research and food analysis). ŻNTJ 3(44):5–22
12. Weist S, Eravci M, Broedel O et al (2008) Results and reliability of protein quantification for two-dimensional gel electrophoresis strongly depend on the type of protein sample and the method employed. Proteomics 8:3389–3396. doi:10.1002/pmic.200800236

Chapter 17

2-DE Separation and Identification of Oat (*Avena sativa* L.) Proteins and Their Prolamin Fractions

Dorota Nałęcz, Iwona Szerszunowicz, Marta Dziuba, and Piotr Minkiewicz

Abstract

At present two-dimensional polyacrylamide gel electrophoresis (2-DE) is the most widely used proteomic tool, which enables simultaneous separation of even thousands of proteins with a high degree of resolution. The quality of 2-DE separation depends on the type of biological material used as a protein source. The presence of interfering compounds (e.g., phenols, as it is the fact in plant material including oat seeds) impedes 2-DE run. With the use of this technique it is possible to analyze the complex protein mixtures, characteristic protein fractions, as well as individual proteins.

The purpose of this chapter is to describe the 2-DE technique (the separate stages of the first and the second dimension) for determining the oat protein composition (oat seed proteome), separation and preliminary identification of oat prolamin fractions. Electrophoretically separated proteins are identified on the basis of p*I* markers (identifying the location of both ends of an IPG strip) and on 2D SDS-PAGE standards. The gel images of oat proteins are analyzed with the help of ImageMaster 2D Platinum 6.0 program (Amersham Bioscience, part of GE Healthcare, Uppsala, Sweden). It allows finding unique spot identifiers for the occurrence of oat prolamin fractions in oat total proteins. The characteristic spots of similar shape and intensity (anchoring spots) and characteristic groups of spots can be searched for the purpose of identification.

Key words Oat proteins, Prolamins, 2-DE separation, Molecular weights, Isoelectric points, Anchoring spots

1 Introduction

Two-dimensional polyacrylamide gel electrophoresis (2-DE) is currently one of the primary tools for proteomics research [1, 2]. It allows simultaneously and most effectively separating a large number (hundreds to thousands) of proteins in two steps. The proteins are first separated on the basis of their isoelectric points (p*I*), the pH at which a protein carries no net charge and will not migrate in an electric field. This first dimension step is called isoelectric focusing (IEF). The proteins separated in the IPG (immobilized pH gradient) strips in the first dimension are applied to the

Sebastian Gasparis (ed.), *Oat: Methods and Protocols*, Methods in Molecular Biology, vol. 1536,
DOI 10.1007/978-1-4939-6682-0_17,

second dimension SDS-PAGE (electrophoresis in polyacrylamide gel in the presence of a sodium dodecyl sulfate (SDS) gels with the use of 2D SDS-PAGE standards [3, 4].

A simultaneous separation of proteins and protein standards in the same gel according to their molecular weights (MW) allows for dependable and reproducible determination of p*I* and MW values of uncharacterized proteins by comparison with selected markers. Proteomic studies of cereal proteins, including their identification, currently focus on wheat proteins [5–7] and rice proteins [8–10], but as yet little research has been done on oat proteins, or the results achieved with the use of different techniques did not bring the expected effects [11, 12].

This chapter describes a standard 2-DE protein separation technique that can be used successfully as a preliminary research method for a proteomic identification procedure of oat total proteins and also oat characteristic prolamin fractions with the help of ImageMaster 2D Platinum 6.0 program (Amersham Bioscience, part of GE Healthcare, Uppsala, Sweden). The program allows for sensitive protein spot detection, gel matching, accurate quality and quantity analysis of 2-DE gels, and statistical analysis (including scatter plots calculation for such parameters as spot intensity, spot area, spot volume to analyze gel similarities and correlation, statistical tests) [13, 14]. It allows finding unique spot identifiers for the occurrence of a given group of proteins [15, 16]. In order to detect prolamins in gels of total seed proteins, gels of oat prolamin fractions can be used as reference gels. To compare spots in the reference gels and in gels of total proteins of oat, the characteristic spots of similar shape and intensity (anchoring spots) and characteristic groups of spots can be searched for the purpose of identification [12].

In this chapter, we show the sample 2-DE gel images of oat total proteins and their prolamin fractions of one selected Flämingstern variety. About 26 spots assigned to the prolamin fractions are present in the gel image of oat prolamins (MW from 19.9 to 25.7 kDa, p*I* from 6.0 to 8.4). The gel image of oat prolamin fractions contains four characteristic series of spots with similar isoelectric points, differing in molecular weights. The organized subregions (groups) of spots can be useful identifiers for the occurrence of oat prolamin fractions in oat total proteins [12].

After the visualization of the separated proteins in gels, they can be then identified with the help of different mass spectrometry (MS) techniques and/or bioinformatics tools (*see* Chapter 18).

2 Materials

2.1 Equipment

2.1.1 Isoelectric Focusing (IEF)

1. Isoelectric Focusing Unit (*see* **Note 1**).
2. Strip holders, 18 cm long.
3. Microcentrifuge.

4. Ultrasonic bath sonicator (*see* **Note 2**).
5. Solvent vacuum microfiltration assembly.

2.1.2 SDS-PAGE

1. Equilibration tubes or serological pipettes (polystyrene) for 18 cm IPG strips.
2. Lab rocker.
3. Gel caster (*see* **Note 3**) and anode assembly cassette carrier (for 25.5 × 20.5 × 0.1 cm gels).
4. Vertical Electrophoresis System for large format gels (*see* **Note 4**).
5. Lab orbital shaker (*see* **Note 5**).
6. Gel electrophoresis software (for analysis and comparison of 2D electrophoresis patterns) (*see* **Note 6**).
7. 2D gel image scanner (for 25.5 × 20.5 × 0.1 cm gels).
8. Solvent vacuum microfiltration assembly.
9. Plastic boxes with lids (for 25.5 × 20.5 × 0.1 cm gels).

2.2 Protein Samples

Oat total protein or their prolamin extract in a liquid/lyophilized form (*see* Chapter 16).

2.3 Reagents

Reagents used for buffer preparation should be of the highest purity (electrophoresis grade). For fixing, staining, and destaining gels, reagents (85 % orthophosphoric acid, methanol) should be pure p.a. To prepare all solutions use deionized water. Filter the buffers after their preparation through a 0.45-μm filter with the use of a solvent vacuum microfiltration assembly to remove any potential residues. Important: Some of the chemicals used for preparation of solutions as Tris(hydroxymethyl)aminomethane (other names: Tris, Tris buffer, Tris base, Trizma, Trisamine), thiourea, dithiothreitol (DTT), acrylamide, *N,N′*-methylenebisacrylamide, *N,N,N′,N′*-tetramethylethylenediamine (TEMED), ammonium persulfate (APS), iodoacetamide (IAA), are very hazardous. Acrylamide is a potent neurotoxin and is absorbed through the skin (the effects are cumulative). Polyacrylamide is considered to be nontoxic, but it should be treated with care, because it may contain small quantities of unpolymerized material. Wear an appropriate laboratory coat, powder-free nitrile gloves, and safety glasses. One should weigh hazardous chemicals in a fume hood, wearing a disposable dust mask.

2.3.1 Isoelectric Focusing (IEF) [3]

1. Immobiline DryStrip gels, 18 cm long, pH 3–10. Store at −20 °C or below (*see* **Note 7**).
2. IPG buffer, pH 3–10 (carrier ampholytes).
3. 1 % (w/v) Bromophenol Blue (BPB) stock solution: to prepare 10 mL of BPB stock solution, dissolve 0.1 g of BPB and 0.06 g Tris base in water, stir, and adjust the volume to 10 mL with water. Store at 4 °C.

4. IPG DryStrip rehydration buffer: 7 M urea, 2 M thiourea, 4% (w/v) 3-[(3-cholamidopropyl)-dimethylammonio]-1-propane sulfonate (CHAPS), 0.5% (v/v) IPG buffer, pH 3–10, 1.8 mM DTT, 0.002% (v/v) BPB. To prepare 25 mL of rehydration stock solution, dissolve 10.5 g of urea (*see* **Note 8**), 3.8 g of thiourea, and 1 g of CHAPS in water, stir, adjust the volume to 25 mL, and filter. Divide all volume on 3 mL aliquots (*see* **Note 9**), and store at −20 °C. Just prior to use, weigh 8.4 mg of DTT, pipette 15 μL of IPG buffer and 6 μL of BPB (*see* **step 3**), and adjust the volume to 3 mL with rehydration stock solution.
5. Immobiline DryStrip cover fluid (*see* **Note 10**). Store at 18–25 °C.
6. Strip holder cleaning solution (GE Healthcare, Fairfield, USA).
7. 96% ethanol.

2.3.2 SDS-PAGE [3]

1. 1.5 M Tris–HCl buffer pH 8.8: to prepare 1 L of 1.5 M Tris–HCl buffer, dissolve 181.7 g of Tris base in 750 mL of water. Adjust pH of the solution to 8.8 with HCl, stir, adjust the volume to 1 L with water, and filter. Store at 4 °C.
2. Equilibration buffer: 6 M urea, 75 mM Tris–HCl pH 8.8, 29.3% (v/v) glycerol (87% w/w), 2% (w/v) SDS, 0.002% BPB. To prepare 250 mL of equilibration buffer, dissolve 90.13 g of urea in water (*see* **Note 8**), add 12.5 mL of 1.5 M Tris–HCl pH 8.8, 86.25 mL glycerol (87% w/w), 5 g of SDS, stir, and adjust the volume to 250 mL. Divide all volume on 60 mL aliquots (*see* **Note 11**), and store at −20 °C. Just prior to use, add 120 μL of to each aliquot.
3. Acrylamide/bis-acrylamide stock solution, 30% T, 2.6% C: 30% acrylamide, 0.8% *N,N'*-methylenebisacrylamide. To prepare 500 mL of acrylamide/bis-acrylamide stock solution, dissolve 150 g of acrylamide and 4 g of *N,N'*-methylenebisacrylamide in ~400 mL of water, adjust the volume to 500 mL with water, and filter. Store at 4 °C in the dark.
4. 10% (w/v) SDS stock solution. Store at room temperature.
5. 15% (w/v) sucrose solution (*see* **Note 12**). Store at 4 °C.
6. 10% (w/v) APS solution. Prepare just prior to use (*see* **Note 13**).
7. *N,N,N',N*-tetramethylethylenediamine (TEMED). Store at room temperature.
8. Homogeneous gel solution: final gel concentration 12.5%. To make 650 mL of the solution (*see* **Note 14**), add 271.05 mL of acrylamide/bis-acrylamide stock solution, 162.5 mL of 1.5 M Tris–HCl buffer pH 8.8, 6.5 mL of SDS, 13 mL of 15% (w/v) sucrose solution, 193.5 mL of water, and after deaeration 3.25 mL of 10% (w/v) APS and 214.5 μL of TEMED (*see* **Note 15**). Prepare just prior to use.

9. Water-saturated butan-1-ol: mix 25 mL of butan-1-ol with 2.5 mL of water. Store at room temperature.
10. 2-D SDS-PAGE standards: prepare application pieces saturated with the standards. To that end cut ~0.7 × 0.7 cm pieces out of filter paper. Then thoroughly mix the standard protein solution, and apply 10–15 μL of the solution to each application piece prepared earlier. The filter paper and prepared application pieces should be held with the forceps. Store at −20 °C (*see* **Note 16**).
11. Electrode buffer stock solution: 97 mM Tris base, 798 mM glycine, 0.4 % (w/v) SDS. To prepare 2 L of electrode buffer, dissolve 23.5 g of Tris base, 119.84 g of glycine and 8 g of SDS in ~1.5 L of water, stir, and adjust the volume to 2 L with water (*see* **Note 17**). Store at room temperature.
    (a) Anode buffer: to prepare 1.5 L of anode buffer dilute a stock solution with water in ratio 1:4.
    (b) Cathode buffer: to prepare 1 L of cathode buffer dilute a stock solution with water in ratio 1:2.
12. 0.5 % (w/v) agarose solution: to make 250 mL of the solution, dissolve 1.25 g of agarose in 125 mL of electrode buffer stock solution and 125 mL of water, in a 500 mL Erlenmeyer flask. Swirl to disperse. Heat in boiling water bath until the agarose is completely dissolved (*see* **Note 18**). Do not allow the solution to boil over. Add a trace of BPB. Store at room temperature.
13. Roti®-Blue (5 × concentrate)-colloidal Coomassie Brilliant Blue G-250 (Carl ROTH, Karlsruhe, Germany).
14. Fixing solution: to prepare 3 L (per 6 gels) of fixing solution, mix 2370 mL of water, 30 mL of 85 % (v/v) orthophosphoric acid and 600 mL of methanol. Prepare solution fresh (*see* **Note 19**).
15. Roti®-Blue gel staining solution: to prepare 3 L (per 6 gels) of staining solution, mix 1800 mL of water and 600 mL of methanol, next stir in 600 mL of Roti®-Blue (5 × concentrate). Prepare solution fresh prior to staining (*see* **Note 20**).
16. Washing (gel destaining) solution: to prepare 3 L (per 6 gels) of destaining solution, mix 750 mL of methanol and 2250 mL of water. Prepare solution fresh prior to destaining (*see* **Note 21**).

3 Methods

All activities (procedures) associated with the preparation and run in the first and the second dimension should be performed in powder-free nitrile gloves.

3.1 Isoelectric Focusing (IEF) of Oat Proteins: The First Dimension

1. Mix the appropriate amount of oat protein extract (*see* **Note 22**) in a liquid or lyophilized form (*see* **Chapter 16**) with IPG DryStrip rehydration buffer in an Eppendorf tube, and vortex.
2. Keep the oat protein sample in an ultrasonic bath sonicator for 5 min (sonicate the samples five times with 30 s bursts and cooling on ice), and leave them for the next 55 min at room temperature (*see* **Note 8**).
3. Add IPG buffer, DTT, and BPB to the oat protein sample containing rehydration buffer, vortex and wait for 1 h.
4. Centrifuge the oat protein sample prior to loading into the strip holder using a microcentrifuge (10,000 ×*g*, for 5 min) to remove insoluble material and particulate matter.
5. Position the required number of strip holders on the leveled platform of Ettan IPGphor. Make sure the sharp, anodic (+) end of each strip holder faces the anode (+).
6. Pipette 350 μL of oat protein sample containing rehydration buffer using a micropipette into IPG DryStrip, 18 cm long each (*see* **Note 23**).
7. Deliver the sample solution slowly as a strip of liquid between the two electrodes of a strip holder, pressing micropipette tips on the holder, and moving it this way to spread the protein sample evenly and to avoid air bubbles. Any large bubbles must be removed.
8. Remove the protective cover foil from Immobiline DryStrip gel taken out of the freezer with forceps, starting at the basic (−) end (*see* **Note 24**).
9. Using the forceps, slowly position Immobiline DryStrip gel, gel side down, in the holder and onto the rehydration buffer solution, lower the gel acidic (anodic) end first. Make sure that the sharp, anodic (+) end of the strip faces the anodic end of the strip holder.
10. To help wet the entire IPG DryStrip, gently lift, and lower the strip and slide it back and forth along the surface of the buffer. Next, lower the cathodic (basic) end of the strip into the holder, making sure that the gel contacts the strip holder electrodes at each end and that no air bubbles are trapped under the strip.
11. Pipette the Immobiline DryStrip cover fluid in the volume of ~0.7 mL and cover with it the whole surface of each strip to minimize evaporation and prevent urea crystallization. If some drop of fluid falls on the walls of the holder, wipe it carefully with a small piece of filter paper.
12. Apply the plastic cover of the holder. Ensure that the strip holders lie evenly on the platform of the Isoelectric Focusing

Unit (you can use a plastic ruler) and that holder electrodes have good contact with the electrodes of the Isoelectric Focusing Unit.

13. Close the safety lid of the apparatus and perform rehydration at room temperature for 13 h in the first step and IEF separation of oat proteins for the next ~12 h in the second, under the following running conditions: (1) rehydration-30 V for 13 h; (2) initial IEF-100 V for 1 h, 300 V for 1 h, 600 V for 1 h, 1000 V for 0.5 h; (3) IEF to the steady-state-gradient from 1000 to 8000 V within 0.5 h, 8000 V to the steady-state (*see* **Note 25**).
14. After the completion of IEF, remove the IPG strips from the holders using the forceps, and wash the strip holders with a strip holder cleaning solution (*see* **Note 26**). Wipe the IPGphor platform with 96% ethanol. You can equilibrate focused IPG strips immediately or store the wrapped IPG strips at −80 °C in glass tubes or serological pipettes overnight (*see* **Note 27**), until further SDS-PAGE analysis.

3.2 SDS-PAGE of Oat Proteins: The Second Dimension

3.2.1 IPG Strip Equilibration

Focused in IEF run IPG strips should be next equilibrated in SDS-containing buffer to prepare the oat protein sample for the second-dimension separation. The equilibration procedure consists of two steps. The equilibration buffer in the first step contains DTT, which reduces denatured proteins. The equilibration buffer in the second step contains IAA, which covalently modifies (alkylates) the reduced sulfhydryl groups. The presence of SDS in the equilibration buffer results in proteins becoming negatively charged for the second dimension run.

1. Prepare an appropriate amount of an equilibration solution with DTT (10 mL per 18 cm long strip and 100 mg of DTT per 10 mL buffer).
2. Prepare an appropriate amount of an equilibration solution with IAA (10 mL per 18 cm long strip and 250 mg of IAA per 10 mL buffer).
3. Take out the glass tubes or serological pipettes containing focused IPG strips from −80 °C freezer, and allow them to warm at room temperature for 10 min. If the focused IPG strips are not frozen, remove them from the strip holders after IEF run and place them in equilibration tubes or serological pipettes.
4. Add 10 mL of DTT-containing equilibration buffer to each glass tube/serological pipette. Cap the tubes or seal the pipettes with flexible parafilm. Make sure the gel side of the strip is facing upward and overlaid with the solution.
5. Place the tubes or pipettes with the strips on a rocker platform (*see* **Note 28**), and equilibrate for 15 min at room temperature.

6. Carefully pour the solution into the sink.
7. Add 10 mL of IAA-containing equilibration buffer instead of DTT-containing equilibration buffer, and repeat **steps 4–6**.
8. Strips are now ready for the second dimension electrophoresis.

3.2.2 Preparation of Lab-Cast Gels

1. In order to prepare polyacrylamide gels manually, one needs to prepare a lab kit for pouring them. Among the elements included in such kit there are factory-prepared glass cassettes (e.g., sized 27.6 × 21.7 × 0.75 cm).
2. Preparing polyacrylamide gels: the prepared gel solution must be quickly but carefully poured to the assembled gel caster (e.g., Ettan DALTsix Gel Caster). Cast the gel by introducing the monomer solution into the gel mold in a steady stream to minimize the introduction of oxygen (*see* **Notes 29** and **30**).
3. Using an automatic pipette apply ~1 mL of butan-1-ol saturated with water (*see* **Notes 31** and **32**) on the surface of the poured gel solution in the glass cassette.
4. Do not disassemble the unit for at least 1.5–2 h, until the gels are polymerized (*see* **Note 33**). During the polymerization of gels start the equilibration of focused IPG strips (as described in Subheading 3.2.1, **steps 1–8**), apply the standards of molecular weights to the application pieces, and prepare the apparatus for electrophoresis in the second dimension.
5. Then disassemble the gel caster unit and take out the glass cassette (*see* **Note 34**).
6. Remove butan-1-ol from the surface of the polymerized gel. Rinse the surface of the gel and the entire glass cassette with deionized water, then dry it with a paper towel (*see* **Note 35**), and slide to the cassette carrier (up to ~½ of the cassette height) (*see* **Note 36**). The shorter plate of the glass cassette must face upward.

3.2.3 Placing the Equilibrated IPG Gel Strips on the Top of the Vertical SDS-PAGE Gels

1. After the equilibration immerse the IPG strip in the stock solution of electrode buffer. Then, holding the end of the strip with forceps, place it on the upper part of the longer plate of the cassette. The IPG strip must be in the middle of the gel (*see* **Note 37**).
2. Place the application piece with applied protein standards behind the IPG strip (from the basic (−) end), in the same way as the IPG strip.
3. Hold the IPG strip or application piece with a ruler or spatula, and press to the gel surface. At the same time start the polymerization of the IPG strip and standards applied to the application piece. Put the molten agarose with an automatic pipette on the entire length of the strip and application piece. The agarose polymerizes within ~2 min.

4. Push the glass cassette with the polymerized IPG strip to the bottom of the cassette carrier with the upper buffer chamber with slots placed on it. Fill all the slots of the cassette carrier with prepared gels together with IPG strips polymerized in the gels and molecular weight standards.
5. Carefully seal the slots of the upper buffer chamber and the cassettes in places where gaskets are located, piling up the agarose solution. When the agarose is polymerized (time: ~10 min) place the anode assembly in the chamber of the electrophoretic apparatus (e.g., Ettan DALTsix Electrophoresis Unit).

3.2.4 SDS-PAGE Run

1. The process of electrophoresis should be performed at 20 °C; therefore, the electrophoretic apparatus must be connected to a thermostat.
2. Pour the anode buffer to the lower chamber of the Electrophoresis Unit (up to ~1/3 of its height), turn on the circulation pump and the thermostat (at 20 °C).
3. Place the carrier with glass cassettes containing gels with polymerized IPG strips and molecular weight standards in the chamber of the apparatus.
4. Fill the chamber of the apparatus first with the anode buffer, and then with the cathode buffer.
5. Place the safety lid on the Electrophoresis Unit, and connect the apparatus to a power supply.
6. Set the separation parameters in step 1: voltage 300 V, current 5 mA per gel (sized 25.5 × 20.5 × 0.1 cm), time ~2 h. Then, when the samples penetrate the polyacrylamide gels, change the separation parameters in step 2: voltage 300 V, current 15 mA per gel, time ~20 h (*see* **Note 38**).
7. After the electrophoresis is completed, turn off the power supply and the thermostat.
8. Take off the safety lid of the Electrophoresis Unit together with leads. Carefully pull the upper buffer chamber with the cassettes (*see* **Note 39**).
9. Open the gel cassettes and take out the gels. Gel should adhere to one of the glass plates. Mark the beginning and end of the IPG strip on the bottom of each gel (sized 25.5 × 20.5 × 0.1 cm). Remove the IPG strip, application piece, and agarose gel from the surface of the glass plate. Gently move the gel to a plastic box with a fixing solution.
10. Fill the lower chamber of the Electrophoresis Unit several times with distilled water, then with deionized water, and rinse every time with the circulation pump turned on (for about 3 × 5 min).

3.2.5 Roti®-Blue Gel Staining and Destaining

1. Keep the gels for 1 h in the boxes with fixing solution on a shaker.
2. Then immerse the gels in a Roti®-Blue solution and store in it for ~16 h on a shaker.
3. After the staining, wash gels to destain the background with the use of a destaining solution. Finally rinse gels with deionized water.

3.2.6 Computer-Aided Image Analysis of Oat Protein Gels

1. Destained oat protein gels are next scanned (*see* **Note 40**) to obtain raw scans (*see* Fig. 1a, b) and subjected to an image analysis with the use of ImageMaster 2D Platinum 6.0 program. The computer-aided image analysis of each oat protein sample should be performed for minimum three replications for statistical purposes.
2. Oat protein gel images can be processed according to GE Healthcare ImageMaster 2D Platinum 6.0 User Manual [13]. Oat protein spots can be detected based on the following parameters selected arbitrarily: minimum area 20; smoothness 10; saliency 150 [7, 14].
3. Manual processing is restricted to remove artifacts recognized as spots. Apparent p*I* is calculated using markers that identify

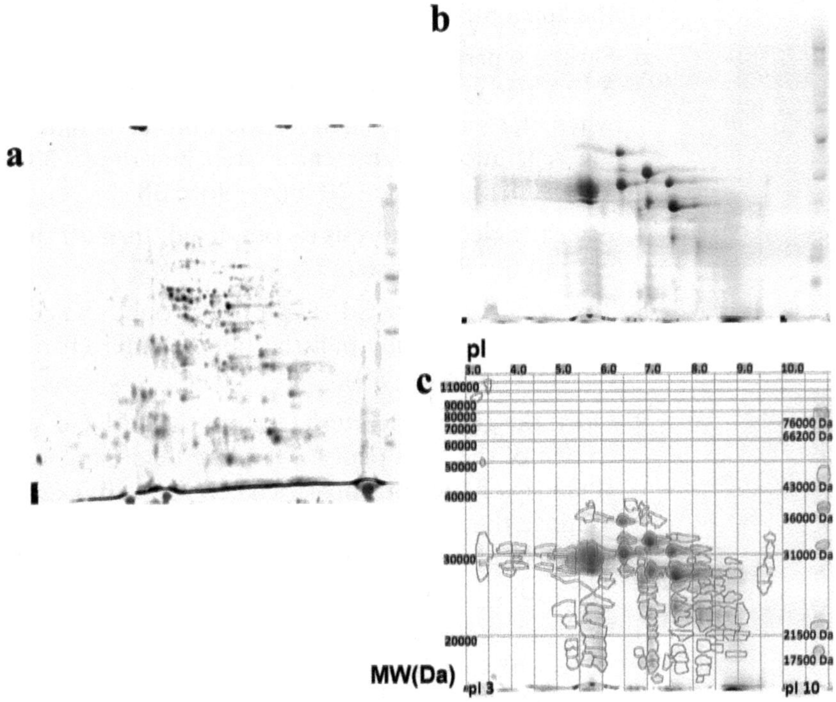

Fig. 1 2-DE images of oat seeds proteins (var. Flämingstern). (**a**) Raw scan of oat total proteins. (**b**) Raw scan of oat prolamins (avenins). (**c**) Oat prolamins image processed with Image Master 2D Platinum software. Molecular weights of standard proteins are marked on the *right*

the location of both ends of an IPG strip. Apparent MW are calculated on the basis of 2D SDS-PAGE standards labeled as shown in Fig. 1c.

4. p*I* markers and MW standards can be used as anchors for matching oat protein gel images. Scatter plots for such parameters as spot intensity, spot area, spot volume can be constructed for matched oat protein spots with the apparent p*I* and MW ranges, to analyze gel similarities, correlation, and statistical tests.

All the above-mentioned opportunities are available as options of the ImageMaster 2D Platinum program [13]. Correlation coefficient of scatter plot may be used as quantitative measure of similarity between two gels or selected fragments of two gels [14]. Another opportunity to measure similarity between two gels or gel fragments is presented in **Note 41**.

4 Notes

1. We use Ettan IPGphor II Isoelectric Focusing Unit (Amersham Bioscience, part of GE Healthcare, Uppsala, Sweden) that allows simultaneous IEF run up to 12 IPG ceramic strip holders from 7 to 24 cm in length or using Ettan IPGphor II Manifold which enables IEF of up to 12 IPG strips. It can be programmed with up to ten (nine step each) user-defined IEF protocols. For any one protocol it is possible to choose and edit any existing protocols, or to make an entirely new protocol to optimize experimental performance.

   Programmable functions include the number of IPG strips, rehydration time, platform temperature, maximum current limit, voltage change pattern, voltage for each step in the isoelectric focusing program, and step duration. Optionally, you can use a more advanced Ettan IPGphor 3 Isoelectric Focusing Unit or other.

2. You need a sonicator for the samples of small volume and reasonable power. A tip sonicator with the possibility of pulse work and regulation of the time of impulses, equipped with tube coolers that keep your sample cold during sonication, will be the best. If the use of it is impossible, you must prevent the protein sample from overheating. Be aware that ultrasonication transmits power into a liquid. Mechanical oscillations lead to turbulence and friction within the liquid. For this reason ultrasonication generates considerable heat during processing. Pulse sonication with short time and effective cooling is needed to reduce the heat-up. For smaller samples, Eppendorf tubes, vials, or glass beaker should be kept in an ice-bath or cooled with ice between bursts for heat dissipation.

3. Some apparatuses for gel casting (e.g., Ettan DALTsix Gel Caster, Amersham Bioscience, part of GE Healthcare, Uppsala, Sweden) include glass cassettes, blank cassettes, separator sheets, filler sheets, among other things. Be sure that the gel caster, glass and blank cassettes, separator, and filler sheets are perfectly clean and dry.
4. We use Ettan DALTsix Large Vertical System (Amersham Bioscience, part of GE Healthcare, Uppsala, Sweden). It allows SDS-PAGE run up to six large format (25.5 × 20.5 × 0.1 or 0.15 cm), lab-cast (you can pour gels in laboratory), or precast gels DALT Gel 12.5 (you can buy ready-made polyacrylamide gels bound to a plastic support film, which provides ease of handling and dimensional stability) simultaneously. It is the best for 18 and 24 cm long IPG strips. Optionally, you can use Ettan DALTtwelve Large Vertical System, which enables SDS-PAGE separation of 12 large format gels or other systems. An example electrophoresis system consists of: an apparatus for electrophoresis (e.g., Ettan DALTsix), power supply (e.g., the EPS 601 power supply), and a thermostat (e.g., the MultiTemp III Thermostatic Circulator).
5. Use an orbital shaker with a shaking platform of adequate dimensions (for six standing large plastic boxes 26 × 21 cm), the anti-skidding rubber mat and the regulator of the time and speed of shaking with the possibility of continuous work (for overnight gel staining). We recommend lab shaker type WL-2000 (JW Electronic, Warsaw, Poland).
6. We use ImageMaster 2D Platinum 6.0 program (Amersham Bioscience, part of GE Healthcare, Uppsala, Sweden). ImageMaster 2D Platinum 6.0 software ensures fast and reliable image comparisons. It easily manages multiple image analyses and offers the possibility to automate detection and spot matching. It enables quantitative and qualitative analysis [13]. In case of the presence of artifacts that are non-protein spots (e.g., bubbles, dust, damages of gel surface) manual processing is restricted to the removal of artifacts recognized as spots.
7. Use 18 cm long strips, with the broad-range pH 3–10, for maximal resolution and loading capacity. It will allow you to detect more oat protein spots and make it easy to select and identify in 2-DE image, but you should provide for a long time of analytical run, in both the first and the second dimension separations (about 25–26 h and 22 h, respectively).
8. Never heat a urea solution above 37 °C (also during sonication) to reduce the risk of protein carbamylation, because urea will degrade to form isocyanate. Isocyanates react with the amino terminus of proteins, preventing N-terminal sequencing. Isocyanates also react with the side chains of lysine residues, resulting in a protein that is unsuitable for many

enzymatic digests. In addition, carbamylation often leads to confusing results from peptides that have unexpected molecular weights. You must be particularly cautious with 2-DE protein samples containing high concentration of urea to avoid heating the sample above 37 °C. Protein modification due to carbamylation will result in artifacts in the image of 2-DE gels.

9. For one experiment 3 mL of rehydration buffer is needed, at the full equipment (6 strips × 0.5 mL of buffer). A given amount of buffer (25 mL) allows for performing eight experiments, at the full mounting of apparatus.
10. Cover fluid is a mineral oil that protects immobiline drystrip gels from drying and from the atmosphere during the first dimension of 2-DE. The necessary volume of mineral oil depends on the length of the IPG strip and the holder and the maximum quantity must allow for covering the holder with a plastic lid.
11. For one experiment you need 60 mL of equilibration buffer with DTT, and 60 mL of equilibration buffer with IAA is needed at the full equipment (6 strips × 10 mL of buffer two times). A given amount of buffer (250 mL) allows for performing two experiments, at the full mounting of apparatus.
12. To minimize breaking of large format gels into pieces during their transfer, we recommend the addition of 2 % sucrose (the final concentration) to the gel solution.
13. Fresh APS crackles when water is added. If it does not, replace it with fresh stock. The old solution will cause that the rate of polymerization, and hence the pore size, will be different.
14. For pouring one homogeneous gel containing 12.5 % of total acrylamide, 100 mL of gel solution is needed.
15. The reaction of gel polymerization is initiated by APS along with TEMED, which accelerates the rate of the formation of free radicals from persulfate, and these in turn catalyze the polymerization. After the addition of TEMED and APS, swirl the solution gently but thoroughly. Mix the initiators completely without introducing too much oxygen. Swirling too little can result in uneven polymerization.
16. 2D SDS-PAGE standards are used for MW determination of sample proteins, monitoring the progress of an electrophoretic run, or as a positive control for analysis conditions. The 2D SDS-PAGE standards used by us are a mixture of seven proteins (MW 17.5–76 kDa, with p*I* 4.5–8.5) (*see* Fig. 1c). The seven reduced, denatured proteins can be visualized with Roti®-Blue stain (*see* **Note 20**). No dilution is required.
17. A given amount of electrophoresis buffer is needed for the preparation of anode and cathode buffers, agarose solution, and rinsing of IPG strips prior to their applying to polyacrylamide gels.

18. Take care to ensure the agarose solution is warm when it is loaded. Do not allow it to solidify. Pipette slowly the molten agarose into the gel cassette to seal the IPG strip to the gel during the second dimension. There must not be any bubbles trapped in the thin agarose layer between the IPG strip and the top of the gel. Plug any gaps with molten agarose if necessary (along the upper buffer chamber with all slots).
19. Per gel (sized 25.5 × 20.5 × 0.1 cm) 0.5 L of fixing solution is needed.
20. Roti®-Blue is a colloidal Coomassie Brilliant Blue G-250 staining solution for proteins in polyacrylamide gels. The sensitivity of Roti®-Blue staining can be compared to that of silver staining (<30 ng/protein) [17]. Store the Roti®-Blue (5 × concentrate) solution at 20 °C. Thoroughly shake the Roti®-Blue (5 × concentrate) solution several times before use. Per gel (sized 25.5 × 20.5 × 0.1 cm) 0.5 L of staining solution is needed.
21. Per gel (sized 25.5 × 20.5 × 0.1 cm) 0.5 L of washing solution is needed.
22. The amount of oat protein extract depends on oat protein concentration. We use 2.5 mg of oat total protein lyophilizate and 8 mg of oat prolamin lyophilizate in 0.5 mL of rehydration buffer, or we add 350 μL of rehydration buffer to 150 μL of the liquid protein extract.
23. We recommend to load 350 μL of rehydration buffer including an oat protein sample per 18 cm strip. The IPG strip should be taken out of the freezer just a few minutes before use to prevent it from taking up water from the environment.
24. Removal from the basic (−) end prevents damage to the acidic (+) end of the gel, which is softer and sensitive, but it depends on strip's producer.
25. Ettan IPGphor running conditions for analytical IEF of oat proteins with the use of Immobiline DryStrip gels, 18 cm long, pH 3–10 are presented according to a typical protocol.

    It is recommended to apply low voltages (30–50 V) during IPG strip rehydration for better entry of high MW proteins into the polyacrylamide gel matrix. Then voltage is stepwise increased up to 8000 V (when IPG strips with separation distances <11 cm are used, voltage should be limited to 5000 V only) [3]. The total time of the IEF run will be dependent on the type of the sample (contents of interfering compounds) and resistances associated with it during IEF separation. We suggest to be guided in the last phase of IEF by the amount of volt-hours (Vh), but not of hours of run to ensure good focusing (we finish IEF run when 60,000 Vhs are achieved).

26. The strip holders are made of aluminum oxide ceramic of minimal thickness for efficient heat transfer and are fragile. Handle the holders with care, wearing powder-free nitrile gloves to avoid contaminating them (prevent skin keratin contamination). Wash each holder after each first dimension IEF run with the supplied strip holder cleaning solution to remove residual protein. Rinse thoroughly with deionized water. The holder must be completely dry before use.
27. The equilibration process can be performed using glass equilibration tubes with screw caps or polystyrene serological pipettes of adequate length for 18 cm long strips (pipettes of 10 mL volume without tips are the best). After positioning a strip in serological pipettes you must seal the pipettes with flexible parafilm.
28. Use a lab rocker with the platform of adequate dimensions, the anti-skidding rubber mat, and the regulator of the time and speed of mixing. If you do not have a lab rocker, you can equilibrate focused IPG strips holding all tubes/pipettes in your hands carefully, with a rocking motion.
29. The apparatus must be filled with a gel solution at such speed so that air bubbles cannot get to glass cassettes. The presence of bubbles in the cassettes will lead to the creation of holes in the polymerized gels. Such gels cannot be used as carriers in the polyacrylamide gel electrophoresis.
30. The apparatus must be filled with a gel solution only up to a certain height. One should remember that after the polymerization and obtaining polyacrylamide gels, strips with separated compounds (after IEF) and application pieces with applied MW standards will be put on gel surfaces.
31. Add water-saturated butan-1-ol to exclude air, and ensure a level surface on the top of the gel.
32. Avoid the contact of butan-1-ol with plastic surfaces of the assembly (e.g., Ettan DALTsix Gel Caster).
33. In case the proteins separated with the 2-DE technique are next to be identified with mass spectrometry, the polyacrylamide gels should be left at 20 °C even for 1 day.
34. From the glass cassettes taken out of the gel caster unit remove the adhering polymerized acrylamide. Rinse the glass cassettes with deionized water. Do the same with the remaining elements of the gel caster unit. Wash the entire unit in water with the addition of a mild detergent (e.g., SDS), then carefully remove the detergent, and rinse the unit with deionized water.
35. Paper towel cannot have color imprint, especially if the next step after 2-DE separation will be analysis using mass spectrometry.

36. Remove the safety lid of the electrophoretic apparatus together with the leads that enable its connection to the power supply. Take out the cassette carrier with the upper buffer chamber. The bottom of the upper chamber appears after the chamber gasket is sealed with agarose solution.
37. In order to put the IPG strip between the two glass plates of a cassette, one should move the plastic part of the strip toward the surface of polymerized gel using a thin plastic spatula or ruler. Pay attention in order not to damage the gel strip. Gel on the IPG strip cannot touch the opposite glass plate. It should adhere at its entire length to the surface of gel in the glass cassette. Between the IPG strip and the surface of gel there cannot be any air bubbles, which interfere with electrophoretic separation.
38. Stop SDS-PAGE run when the Bromophenol Blue dye front line is approximately 1 cm away from the bottom of the gel. To be sure you can designate the stop line in the front of your lower buffer chamber (tank).
39. Pull partly upward the upper buffer chamber checking if the cassettes remain in the anode assembly, which should be kept in a bowl. Take out completely the upper buffer chamber, and then individual gel cassettes from the anode assembly. Rinse all components of the apparatus with distilled water, then with deionized water.
40. Put gels on the scanner platform in such a way that there are no air bubbles between the gel and the surface of scanner.
41. Parameter S describing similarity between two gels or between selected groups of spots at different gels may be expressed using Eq. 1:

$$S = r \times n_{\text{match}} / n \qquad (1)$$

r-correlation coefficient of scatter plot

n_{match}-number of spots matched between two gels or selected areas at two gels

n-number of spots in gel or gel fragment containing smaller number of spots

S value for two identical gels is 1.

Acknowledgment

This work was supported by the Ministry of Science and Higher Education, Poland within project No NN 312 286333.

References

1. Rabilloud T, Chevallet M, Luche S et al (2010) Two-dimensional gel electrophoresis in proteomics: past, present and future. J Proteom 73:2064–2077
2. Rogowska-Wrzesinska A, Le Bihan M-C, Thaysen-Andersen M et al (2013) 2D gels still have a niche in proteomics. J Proteom 88:4–13
3. Görg A (2004) 2-D electrophoresis. Principles and methods. Handbooks from GE Healthcare. GE Healthcare, Uppsala
4. Bio-Rad (2015) 2-D electrophoresis for proteomics: a methods and products manual. http://www.bio-rad.com/LifeScience/pdf/Bulletin_2651.pdf. Accessed 10 Aug 2015
5. Mamone G, Addeo F, Chianese L et al (2005) Characterization of wheat gliadin proteins by combined two-dimensional gel electrophoresis and tandem mass spectrometry. Proteomics 5(11):2859–2865
6. Guo G, Lv D, Yan X et al (2012) Proteome characterization of developing grains in bread wheat cultivars (*Triticum aestivum* L.). BMC PlantBiol12(147):2–24.doi:10.1186/1471-2229-12-147
7. Dziuba M, Nałęcz D, Szerszunowicz I et al (2014) Proteomic analysis of wheat α/A- and β-gliadins. Czech J Food Sci 32(5):437–442
8. Komatsu S, Yano H (2006) Update and challenges on proteomics in rice. Proteomics 6(14):4057–4068
9. Hashimoto M, Komatsu S (2007) Proteomic analysis of rice seedlings during cold stress. Proteomics 7(8):1293–1302. doi:10.1002/pmic.200600921
10. Yang Y, Dai L, Xia H et al (2013) Protein profile of rice (*Oryza sativa*) seeds. Genet Mol Biol 36(1):87–92
11. Jacobsen S, Nesic L, Petersen MK et al (2001) Classification of wheat varieties: use of two-dimensional gel electrophoresis for varieties that cannot be classified by matrix assisted laser desorption/ionization-time of flight-mass spectrometry and an artificial neural network. Electrophoresis 22(6):1242–1245
12. Nałęcz D, Dziuba J, Minkiewicz P et al (2009) Identification of oat (*Avena sativa*) and buckwheat (*Fagopyrum esculentum*) proteins and their prolamin fractions using two-dimensional polyacrylamide gel electrophoresis. Eur Food Res Technol 230:71–78
13. GE Healthcare (2015) ImageMaster 2D Platinum 6.0 User Manual. http://www.genebio.com/_library/pdf/GE_DF_ImageMaster_v2.pdf. Accessed 5 Aug 2015
14. Dziuba J, Minkiewicz P, Nałęcz D et al (2009) Reproducibility of two-dimensional electrophoresis gel images of pea (*Pisum sativum* L.) seed proteins evaluated using scatter plots-a short report. Pol J Food Nutr Sci 59(2):143–146
15. Pietrogrande MC, Marchetti N, Dondi F et al (2002) Spot overlapping in two-dimensional polyacrylamide gel electrophoresis separations: a statistical study of complex protein maps. Electrophoresis 23:283–291
16. Pietrogrande MC, Marchetti N, Cecconi D et al (2003) Spot overlapping in two-dimensional polyacrylamide gel electrophoresis maps: relevance to proteomics. Electrophoresis 24:217–224
17. Carl ROTH (2015) Instructions for use. Roti®-Blue Colloidal Coomassie staining. https://www.carlroth.com/downloads/ba/en/A/BA_A152_EN.pdf. Accessed 3 Aug 2015

Chapter 18

Selected Bioinformatic Tools and MS (MALDI-TOF, PMF) Techniques Used in the Strategy for the Identification of Oat Proteins After 2-DE

Iwona Szerszunowicz, Dorota Nałęcz, and Marta Dziuba

Abstract

Computer analysis of protein maps obtained from the separation of proteins with two-dimensional polyacrylamide gel electrophoresis (2-DE), in combination with mass spectrometry (MS) analysis and selected bioinformatic tools is used in the strategy for the identification of oat proteins. In proteomic research the most often used MS technique is the combination of ion sources: matrix-assisted laser desorption/ionization (MALDI) and the analyzer of the time of flight (TOF), i.e., MALDI-TOF MS.

This chapter describes the possibilities of the use of selected bioinformatic tools (UniProtKB database, ProtParam, Compute p*I*/MW programs) for initial identification of separated oat proteins (especially prolamin fractions) with the 2-DE technique. Also the procedure of preparation of samples obtained from cut out protein spots for analysis with the MALDI-TOF MS and peptide mass fingerprinting (PMF) technique is presented.

Among oat prolamins separated with the 2-DE technique (*see* Chapter 17), 13 protein spots are considered to be the most characteristic (range of MW 27.0–34.6 kDa, p*I* 5.7–7.6) for this fraction of proteins. Among them there are four protein spots (MW 27.0–28.0 kDa) and two spots (MW 31.4–32.1 kDa) which can correspond to avenins (Accession numbers (AC) in UniProtKB: L0L5I0, I4EP88, I4EP64, L0L4I8 and F2Q9W5, L0L6J0, respectively).

Key words Oat proteins, UniProtKB, ProtParam, Compute p*I*/MW, 2-DE, MALDI-TOF MS, PMF, Mascot

1 Introduction

At present in the strategy of proteomic research the two-dimensional polyacrylamide gel electrophoresis (2-DE) technique, as well as mass spectrometry (MS) are used in combination with selected bioinformatic tools. Both the 2-DE and MS technique can be used independently or jointly. First, all isolated proteins, e.g., oat proteins or avenins (*see* Chapter 16), are separated with 2-DE (*see* Chapter 17). Then the proteins separated on polyacrylamide gels are stained for visualization and in order to obtain so-called

Sebastian Gasparis (ed.), *Oat: Methods and Protocols*, Methods in Molecular Biology, vol. 1536,
DOI 10.1007/978-1-4939-6682-0_18, © Springer Science+Business Media LLC 2017

proteomic maps (in case of separation of all protein compounds coded by a genome) or protein maps (e.g., avenins) (*see* Chapter 17). These protein maps, i.e., electrophoretic images with protein spots shown, are created after the focusing of proteins (after isoelectric focusing-IEF) followed by separation of proteins in a polyacrylamide gel (electrophoresis in a polyacrylamide gel in the presence of a sodium dodecyl sulfate, SDS-PAGE) together with protein markers (standards) of molecular weights (MW). Appropriate computer software makes it possible to apply a so-called grid with marked values of isoelectric points (p*I*) and MW to obtained protein maps (*see* Fig. 1 in Chapter 17). Then, using the amino acid sequences of oat proteins and their fragments deposited in various protein databases, e.g., the Universal Protein Resource Knowledgebase (UniProtKB) [1], and appropriate computer programs available: ProtParam [2–4], Compute p*I*/MW [5–7], one can compute the theoretical values of p*I* and MW of oat proteins. In this way the computed (theoretical) p*I* and MW can be compared with obtained values of p*I* and MW read off protein maps, thus enabling initial identification of oat proteins separated for example with the 2-DE technique (*see* Chapter 17).

At present, the UniProtKB database (http://www.uniprot.org) [1] is the biggest worldwide source of data concerning amino acid sequences of various proteins and protein fragments. The information contained in it comes from two databases: Swiss-Prot and TrEMBL (Translated EMBL (European Molecular Biology Laboratory) Nucleotide Sequence Data Library). In the database there are annotations concerning the amino acid sequences, taxonomy, function of proteins, posttranslational modifications, and bibliography, among other things. The UniProtKB database is included in the UniProt databases – the biggest in the world and the most comprehensive catalog of information about proteins. UniProt is a central repository of protein sequence and function created by combining the information contained in Swiss-Prot, TrEMBL and Protein Information Resource (PIR) [8–17]. UniProtKB is also one of popular resources of the worldwide bioinformatic portal SIB ExPASy (Expert Protein Analysis System) (http://www.expasy.org) [18] operated by the Swiss Institute of Bioinformatics (SIB). SIB ExPASy provides access also to scientific databases and program tools from many fields and disciplines of science, including proteomics.

In order to perform further identification of proteins separated with the 2-DE technique, one can cut out individual protein spots from the polyacrylamide gel. Having prepared samples from them, one can perform MS analysis, obtaining the mass spectrum of the analyzed protein compound as a result. MS is an analytic technique based on the measurement of a mass-to-charge ratio of a given ion (m/z) [19, 20].

In proteomic research the most often used MS technique is the combination of ion sources: matrix-assisted laser desorption/ionization (MALDI) [21–24] and the analyzer of the time of flight (TOF), i.e., MALDI-TOF MS [19, 20, 25–27]. In this MS technique a mild ionization method is used, where the energy of the laser beam is selected in a way which prevents the molecule fragmentation, and enables to strike the molecules out of a specifically prepared matrix. The matrix absorbs the energy of the laser, which is later transferred to the analyzed molecules [21].

For better protein identification the peptide mass fingerprinting (PMF) technique is also commonly used [19, 20, 28, 29]. A protein spot extracted from gel is subject to enzymatic hydrolysis (digestion), and the mixture of peptides obtained as a result is then analyzed with the use of a mass spectrometer, usually MALDI-TOF. The analyzer of mass measures the molecular mass of obtained peptides on the basis of their m/z value. The obtained mass spectrum is a set of peptide masses included in one protein. In this way a unique protein mass identifier known as a peptide map of a protein or a "peptide mass fingerprint" is created. The basic enzyme used for experimental hydrolysis of a protein is trypsin [19, 20, 28, 29].

In the PMF technique the next step is to compare the obtained protein mass identifier (a set of experimental peptide masses observed in the spectrum) with masses of peptides obtained as a result of theoretical digestion (in silico hydrolysis). In order to perform in silico hydrolysis, the amino acid sequences of proteins deposited in protein databases are used. Furthermore, the same enzyme as the one used at the experimental stage is selected. As a result of in silico hydrolysis, a theoretical set of amino acid sequences, thus a theoretical set of peptide masses obtained as a result of hydrolysis of specific peptide bonds by a selected enzyme, is obtained. Then, with the use of computer algorithms, it is possible to match experimental peptide masses to masses of peptides obtained from in silico hydrolysis. Most often, the program used for protein identification is Mascot. Mascot uses a probabilistic scoring algorithm for protein identification [29, 30], and is available to use for free on the Matrix Science website (http://www.matrixscience.com/search_form_select.html) [31].

This chapter describes the possibilities of the use of selected bioinformatic tools (UniProtKB database [1], ProtParam/Compute p*I*/MW program [2–7]) for initial identification of separated oat proteins (especially prolamin fractions) with the 2-DE technique. Also the procedure of preparation of samples obtained from cut out protein spots for analysis with the MALDI-TOF MS [32] and PMF [33, 34] technique is presented.

2 Materials

2.1 Equipment

1. The UniProtKB database [1].
2. Mass spectrometer of MALDI-TOF type.
3. Ultrasonic bath sonicator.
4. Centrifugal evaporator.
5. Mascot software (Matrix Science, UK) [30, 31].
6. National Center for Biotechnology Information (NCBI) database [35] or Swiss-Prot database [1].

2.2 Amino Acid Sequences of Oat Proteins

In the UniProtKB database there are 266 amino acid sequences of oat proteins (*Avena sativa*) including protein fragments available (entry on 20.11.2015). All amino acid sequences are written in the form of a one-letter amino acid code (*see* **Notes 1** and **2**). Among the 266 sequences, 125 sequences (including sequence fragments) correspond to various proteins (e.g., seed storage proteins, phytochrome, ubiquitin, and other proteins), 97 are sequences of various enzymatic proteins, while 44 sequences (including sequence fragments) correspond to avenins and gliadin-like avenins (*see* Subheading 3.1).

2.3 Polyacrylamide Gels (See Chapter 17)

1. Polyacrylamide gels with 2D electrophoretically separated oat total proteins.
2. Polyacrylamide gels with 2D electrophoretically separated oat avenins.

2.4 Reagents

All reagents used should be of the highest purity (HPLC grade). Wear appropriate laboratory coat, powder-free non latex gloves and safety glasses (*see* **Notes 3–5**).

1. Buffer A: 50% (v/v) acetonitrile (ACN), 0.1% (v/v) trifluoroacetic acid (TFA). To prepare 50 mL of buffer A, mix 24.5 mL of deionized water, 25 mL of ACN, and 0.5 mL of 10% TFA (*see* **Note 6**).
2. 100 mM ammonium bicarbonate (AmBi)/50% (v/v) ACN: to prepare 100 mL of 100 mM AmBi/50% ACN dissolve 0.79 g of AmBi in 40 mL of deionized water, add 50 mL of 100% ACN, and adjust the volume to 100 mL. Store in refrigerator. Use within 1 month.
3. 50 mM AmBi: to prepare 100 mL of 50 mM AmBi dissolve 0.395 g of AmBi in 90 mL of deionized water, and adjust volume to 100 mL.
4. 50 mM acetic acid (AA).
5. Trypsin digestion solution (e.g., trypsin Gold, Promega, Madison, USA): to prepare trypsin digestion solution resuspend

1 vial in 200 μL of the attached resuspension solution (50 mM AA), and then dilute in 800 μL of 50 mM AmBi (to 20 μg/mL). Usually trypsin digestion solution is used for protein identification by peptide mass fingerprinting (PMF) (*see* **Notes 7–9**).

6. Alpha-cyano-4-hydroxycinnamic acid (ACHCA) matrix: to prepare ACHCA matrix (*see* **Notes 10** and **11**) solution use only recrystallized matrix. Add 0.5 mL of buffer A (*see* **Note 12**) to one tube (5 mg ACHCA) to the final concentration of 10 mg/mL. Vortex until everything is dissolved. Spin down for a while. Transfer 0.4 mL to a new tube. To prepare working matrix solution mix 0.1 mL of above solution with 0.3 mL of buffer A (2.25 mg/mL), and vortex quickly. Matrix solution is ready to use and should be stored wrapped in aluminum foil and used within 12 h.
7. 200 mM AmBi/40% (v/v) ACN: to prepare 100 mL of 200 mM AmBi/40% ACN dissolve 1.58 g of AmBi in 50 mL of deionized water, add 40 mL of 100% ACN, and adjust the volume to 100 mL. Store in refrigerator and use within 1 month.
8. Eluting mixture: 50% (v/v) formic acid (FA), 25% (v/v) ACN, 15% (v/v) propan-2-ol, 10% (v/v) deionized water. To prepare 100 mL of eluting mixture mix 25 mL of ACN, 15 mL of propan-2-ol, 10 mL of deionized water with 50 mL of FA. Store in refrigerator and use within 1 month.
9. Buffer B: 50% (v/v) ACN, 0.9% (v/v) TFA. To prepare 50 mL of buffer B, mix 20.5 mL of deionized water, 25 mL of ACN, and 4.5 mL of 10% TFA (*see* **Note 6**).
10. Sinapinic acid (SinA) matrix (*see* **Notes 10** and **11**): vortex 20 mg of SinA with 1 mL of buffer B (*see* **Note 12**) in an Eppendorf tube for 2 min. Spin down in a centrifuge briefly. Transfer the supernatant to a new Eppendorf tube (*see* **Note 13**).

3 Methods

3.1 Selected Bio Informatic Tools Used for Initial Identification of Oat Proteins Separated with the 2-DE Technique (See Note 14)

Oat proteins and protein fragments isolated from seeds (*see* Chapter 16), and then separated with the 2-DE technique (*see* Chapter 17 and Table 1) can be initially identified by using amino acid sequences of proteins available in UniProtKB and bioinformatic tools, e.g., the ProtParam and Compute p*I*/MW programs [2–7, 36] (*see* Table 2).

At present, in the resources of the UniProtKB database there are 266 amino acid sequences of oat proteins and protein fragments, including 44 sequences which belong to avenins and gliadin-like avenins (entry on 20.11.2015) (*see* Subheading 2.2). Among 44 records (lines) obtained for oat prolamin fractions, after opening one of them one gains access to information about a

Table 1
Characteristic spots of proteins separated with 2-DE, present in the image of oat prolamin fractions ($n^a = 3$), (*see* Fig. 1 in Chapter 17), presented according to the diminishing values of molecular weights of protein spots

No	Observed MW value (average value ± SEM[b]) (kDa)	Observed p*I* value (average value ± SEM[b])	No	Observed MW value (average value ± SEM[b]) (kDa)	Observed p*I* value (average value ± SEM[b])
1.	34.6 (±0.1)	6.35 (±0.07)	8.	28.5 (±0.0)	5.70 (±0.08)
2.	32.1 (±0.0)	6.40 (±0.06)	9.	28.5 (±0.3)	7.64 (±0.06)
3.	31.4 (±0.1)	6.94 (±0.07)	10.	28.0 (±0.2)	7.14 (±0.07)
4.	30.4 (±0.0)	5.68 (±0.06)	11.	27.9 (±0.1)	5.71 (±0.04)
5.	30.1 (±0.1)	7.44 (±0.06)	12.	27.1 (±0.2)	7.10 (±0.07)
6.	29.9 (±0.0)	6.37 (±0.07)	13.	27.0 (±0.3)	7.63 (±0.06)
7.	29.1 (±0.0)	5.70 (±0.09)			

The images of gels with separated oat prolamins are presented in Fig. 1 (*see* Chapter 17). It is an image of gel after scanning (*see* Fig. 1b in Chapter 17), with a marked grid of isoelectric points (p*I*) and molecular weights (MW) (*see* Fig. 1c in Chapter 17). The above table presents 13 of the most characteristic spots which appear in the image of oat prolamins separated with the 2-DE technique. A protein spot is considered to be characteristic if it appears on each of the three analyzed gels (number of repetitions performed for isolated oat prolamins ($n = 3$)), and its intensity was the highest in comparison with other spots present on the gel

[a]Number of repetitions

[b]Standard error of the mean

selected avenin or gliadin-like avenin. Among accessible information there is an amino acid protein sequence (*see* **Notes 1**, **15–20**), and next to it there are programs available (*see* **Note 15**), which a database user can choose.

1. ProtParam [2–4] is a tool which allows the computation of various physical and chemical parameters of a protein or its fragment on the basis of an amino acid sequence. The parameters which can be computed by the program are: theoretical MW and p*I* (calculated by Compute p*I*/MW) [5–7] (*see* **Notes 21** and **22**), amino acid composition, atomic composition, extinction coefficient, estimated half-life, instability index, aliphatic index, and the grand average of hydropathicity (GRAVY).

   Table 2 presents theoretical values of p*I* and MW computed with the use of the ProtParam program for 44 sequences of proteins and protein fragments which belong to oat prolamins and are deposited in UniProKB [1]. The MW and p*I* ranges which correspond to 42 oat proteins and protein fragments were (32.79–13.89 kDa) and (4.65–8.97), respectively. In two cases the amino acid sequences of protein fragments (AC Q09097 and Q09095) consisted only of 43 and 36 amino acid residues, respectively.

Table 2
Selected physicochemical data concerning avenins, which come from the resources of the UniProtKB (Swiss-Prot[1] and TrEMBL[2]) database, presented according to the diminishing values of molecular weights of proteins and protein fragments

No	Entry[1 or 2] (AC[a])	Protein names[1 or 2]	Sequence status	Length (AA[b])	MW[c] (kDa)	p*I*[d]
1.	L0L6J0[2]	Gliadin-like avenin[2]	Complete	281	32.79	6.69
2.	F2Q9W5[2]	Avenin protein[2]	Fragment	271	31.61	6.08
3.	L0L4I8[2]	Gliadin-like avenin[2]	Complete	250	29.06	8.16
4.	I4EP64[2]	Avenin[2]	Fragment	243	27.95	6.68
5.	I4EP88[2]	Avenin[2]	Complete	242	27.84	8.15
6.	L0L5I0[2]	Gliadin-like avenin[2]	Complete	229	26.48	6.68
7.	F4MJY2[2]	Avenin protein[2]	Fragment	225	26.01	5.66
8.	L0L5G8[2]	Gliadin-like avenin[2]	Complete	223	25.75	6.69
9.	F2Q9W3[2]	Avenin protein[2]	Fragment	222	25.73	7.73
10.	F4MJY1[2]	Avenin protein[2]	Fragment	221	25.60	5.56
11.	Q09072[2]	Avenin[2]	Complete	222	25.47	8.15
12.	P80356[1]	Avenin-3[1]	Complete	220	25.28	6.68
13.	F2Q9W4[2]	Avenin protein[2]	Fragment	214	24.63	5.66
14.	L0L5H5[2]	Gliadin-like avenin[2]	Complete	214	24.50	7.53
15.	L0L4J7[2]	Gliadin-like avenin[2]	Complete	212	24.50	6.69
16.	I4EP78[2]	Avenin[2]	Complete	212	24.29	8.81
17.	Q2EPY2[2]	Avenin[2]	Complete	212	24.26	6.68
18.	P27919[1]	Avenin[1]	Complete	214	24.23	7.54
19.	L0L6J5[2]	Gliadin-like avenin[2]	Complete	211	24.15	6.68
20.	L0L845[2]	Gliadin-like avenin[2]	Complete	209	24.07	7.54
21.	L0L6K1[2]	Gliadin-like avenin[2]	Complete	208	24.01	8.15
22.	L0L4J1[2]	Gliadin-like avenin[2]	Complete	209	23.99	7.54
23.	I4EP85[2]	Avenin[2]	Complete	209	23.92	8.81
24.	I4EP86[2]	Avenin[2]	Complete	209	23.89	8.80
25.	F4MJY6[2]	Avenin protein[2]	Fragment	202	23.25	8.70
26.	F4MJY3[2]	Avenin protein[2]	Fragment	204	23.23	6.06
27.	L0L6K5[2]	Gliadin-like avenin[2]	Complete	201	23.12	8.15
28.	L0L841[2]	Gliadin-like avenin[2]	Complete	200	22.97	8.15
29.	L0L8A9[2]	Gliadin-like avenin[2]	Complete	200	22.96	8.15
30.	L0L8B6[2]	Gliadin-like avenin[2]	Complete	199	22.87	8.15

(continued)

Table 2
(continued)

No	Entry[1 or 2] (AC[a])	Protein names[1 or 2]	Sequence status	Length (AA[b])	MW[c] (kDa)	pI[d]
31.	F4MJY0[2]	Avenin protein[2]	Fragment	198	22.82	6.87
32.	F4MJY5[2]	Avenin protein[2]	Fragment	198	22.77	8.70
33.	F4MJY4[2]	Avenin protein[2]	Fragment	187	21.53	6.87
34.	L0L833[2]	Gliadin-like avenin[2]	Complete	189	21.46	8.82
35.	L0L837[2]	Gliadin-like avenin[2]	Complete	189	21.42	8.82
36.	L0L8A4[2]	Gliadin-like avenin[2]	Complete	189	21.42	8.82
37.	L0L8A0[2]	Gliadin-like avenin[2]	Complete	188	21.34	8.82
38.	Q09114[1]	Avenin-E[1]	Complete	182	21.04	7.44
39.	L0L5H3[2]	Gliadin-like avenin[2]	Complete	181	20.44	8.97
40.	L0L6J7[2]	Gliadin-like avenin[2]	Complete	181	20.42	8.82
41.	Q09071[2]	Avenin[2]	Complete	181	20.41	8.82
42.	L0L4J5[2]	Gliadin-like avenin[2]	Complete	118	13.89	4.65
43.	Q09097[1]	Avenin-F[1]	Fragment	43	5.21	3.45
44.	Q09095[1]	Avenin-A[1]	Fragment	36	4.39	4.25

MW and p*I* were computed on the basis of an available amino acid sequence from the UniProtKB database and the ProtParam program (including Compute p*I*/MW)
[a]Accession number
[b]Number of amino acids
[c]Molecular weight
[d]Theoretical isoelectric point

In the gel images of oat prolamins separated with the use of the 2-DE technique there were 13 protein spots of the highest intensity (*see* Fig. 1 in Chapter 17). These spots occurred in the MW and p*I* range respectively (MW 27.0–34.6 kDa, p*I* 5.7–7.6). Among them there were four protein spots (MW 27.0–28.0 kDa) and two spots (MW 31.4–32.1 kDa). They can correspond in turn to avenins (AC: L0L5I0, I4EP88, I4EP64, L0L4I8 and F2Q9W5, L0L6J0, respectively). For further identification these most characteristic spots must be cut out of the gel and appropriately prepared for the MS analysis (*see* Subheading 3.2).

2. Among bioinformatic tools available via UniProtKB [1] or ExPASy [18] there is the PeptideMass program [37–39]. This program (*see* **Notes 18–20**) allows to calculate the masses of peptides (*see* **Notes 23–28**) obtained as a result of hydrolysis of a protein or its fragment by a selected enzyme (alternatively

by a chemical compound). It generates information about the masses of obtained peptides, the location of these peptides in the polypeptide chain of the analyzed protein, the amino acid sequence of peptides.

3.2 Identification and Characterization of Oat Proteins by Mass Spectrometry After 2-DE

Use protective clothes and powder-free, non-latex gloves during all routines. Use protein low-binding laboratory plastics (tubes, tips) (*see* **Note 29**).

3.2.1 In-Gel Digestion with Trypsin for Protein Identification by Mass Spectrometry (See Notes 30–32)

1. Cut out protein spots of interest from 2-DE gel (*see* **Notes 33–35**) with a clean scalpel (*see* **Note 36**) and place them in low-binding 0.5 mL tubes.
2. Destain the gel spots twice (or more if gel particles are not clear) with 0.2 mL of 100 mM AmBi/50% ACN for 30 min at 37 °C. Vortex every 10 min. Remove colored liquid with a pipette.
3. Dehydrate gel spots for 5 min at room temperature in 0.1 mL of 100% ACN. Gel pieces become white and shrink. Remove ACN with a pipette and dry gel pieces at room temperature, or use a centrifugal evaporator for 15 min.
4. Rehydrate gel pieces with 20 μL of trypsin (*see* **Notes** 7–9) solution (or volume enough to cover a gel spot) at room temperature for 1 h. Add 10 μL of 50 mM AmBi to keep pieces wet during cleaving.
5. Proceed in gel digestion overnight at 37 °C.
6. Spin down samples by centrifugation at 18,500 × *g* for 1 min. Transfer the supernatant containing tryptic peptides to a new low-binding tube.
7. Add 50 μL of buffer A to gel pieces and sonicate in ultrasonic bath for 10 min. Spin down as previously and transfer the supernatant to the tube from **step 6**.
8. Dry the extracted tryptic peptides in the centrifugal evaporator until they are almost dry. Tryptic peptides solutions are ready for mass spectrometric analysis (*see* **Notes 37–39**), or can be stored at −20 °C.

3.2.2 MALDI-TOF Analysis of Oat Protein Tryptic Digests

1. Add 5 μL of buffer A to tryptic peptide solution and sonicate for 2 min.
2. Premix 1 μL of tryptic peptide solution with 1 μL of ACHCA matrix (*see* **Notes 10** and **11**) solution in a 0.5 mL tube by pipetting up and down five times. Spot immediately on the MALDI target plate. Allow spots to dry completely. Load the plate into the spectrometer.

3. Obtain mass spectra for PMF-based identification using the MALDI-TOF mass spectrometer in positive polarity, using the instrument in reflectron mode with an acceleration voltage of 20 kV and delay extraction of 300 ns. Obtain spectra in automatic mode in mass range *m/z* 800–2500. For calibration use trypsin autolytic fragment ions 842 and 2211 (*m/z*) [19].

3.2.3 Protein Identification in Mascot Server

1. Perform protein identification using the spectra generated by a mass spectrometer, and to identify the protein spot use Mascot software (Matrix Science, UK) [30, 31] (*see* **Note 40**) searching the NCBI [35] nonredundant database or the Swiss-Prot database [1].
2. Proteins are identified by comparison of tryptic peptide product ion mass spectra with those generated from the database. Use carboxymethylation as modification of cysteine for database searching. Protein candidates with program scores >79 are classified as identified (*see* **Notes 41–44**).

3.2.4 Passive Elution of Intact Proteins

1. Carefully cut out spots of interest from the gel using a scalpel (*see* **Notes 33–36**), put each one into a separate 0.5 mL low-binding tube (*see* **Note 45**), and treat with 100 μL of 200 mM AmBi/40% ACN for 30 min at 30 °C. Vortex every 10 min.
2. Repeat two or three times for full destaining. Incubate the destained gel spots in 40 μL of eluting mixture for 10 min in an ultrasonic bath.
3. Centrifuge at 1,500 × *g* for 2 min. Use supernatants for molecular mass analysis.

3.2.5 MALDI-TOF Analysis of the Intact Oat Proteins

1. Mix 1 μL of supernatant with 2 μL of SinA matrix (*see* **Notes 10, 11** and **13**) solution in a 0.5 mL tube with a pipette tip, and spot 0.3 μL of it on the MALDI target plate.
2. Let it dry completely. Load the MALDI plate into the instrument.
3. Record the spectra with the MALDI-TOF instrument equipped with a pulsed nitrogen laser (337 nm, pulse width 5 ns), and operate in delayed extraction linear mode, in positive polarity, with an acceleration voltage of 20 kV. For external calibration use bovine serum albumin (4 mg/mL) [19].

4 Notes

1. In the UniProtKB database amino acid sequences are written in the form of a one-letter code according to the valid recommendations of IUPAC-IUB Joint Commission on Biochemical Nomenclature (JCBN).

2. The UniProt database is updated every 4 weeks.
3. MS detects compounds in femtomole or even attomole amounts of material, therefore it is essential for the samples prepared for MS analysis not to be contaminated with keratins. Keratins, being a component of human skin epithelium and hair, can easily penetrate into a sample. Then they are detected during the analysis with a mass spectrometer as an additional, undesirable mass [29].
4. MALDI-TOF MS is a very sensitive technique which allows to detect even small amounts of impurities in an analyzed sample. Such impurities include: plasticizers, initiators of polymerization released from colored tips, low quality plastic test tubes, and powder from latex gloves. In mass spectra there will be also peaks resulting from carelessly rinsed out detergents used for cleaning glass or sodium chloride. Plastic products used in a laboratory should have higher resistance, especially to organic solvents. Acetonitrile can release from them compounds which interfere with the result of MS analysis.
5. In order to avoid contamination of bacterial origin, it is recommended to prepare solutions before the MS analysis [29].
6. Use glass pipettes when handling TFA with concentration (5–100%) to avoid release of softeners from plastic equipment, which can affect MALDI spectrum. Buffers A and B should be slit to 1 mL of aliquots and store at −20 °C and used within 1 month.
7. Because the mass of a peptide can be reflected in the spectrum if a proton is added to it, trypsin is the most frequently used enzyme. The specificity of activity of this enzyme (it hydrolyzes peptide bonds formed with the participation of lysine or arginine residues from C-terminal end) enables the creation of spots for proton addition. Trypsin is also an enzyme used to conduct digestion of proteins both in solutions and gels.
8. The method of preparation and storage of trypsin is significant for the quality of obtained results of protein identification with the MALDI-TOF technique [29].
9. If there are too few lysine or arginine residues in amino acid sequences of proteins, such enzymes as: V-8 protease, chymotrypsin, or other proteases, e.g., Lys-C, Asp-N, Glu-C [29, 40] are used. Be careful and use only proteomic grade enzymes, and follow the producer's instruction.
10. The selection of matrix depends on analyzed samples (proteins, peptides). ACHCA is a matrix used for peptides and smaller proteins ≤5000 Da. Intense signals for proteins and peptides. No adduct peaks. Protein ions may undergo metastable decay at very low laser energies and low protein concentration. We recommend ACHCA from LaserBio Labs, whereas sinapinic acid (SinA)

(other names: 3-(4-hydroxy-3,5-dimethoxyphenyl)prop-2-enoic acid, sinapic acid, 3,5-dimethoxy-4-hydroxycinnamic acid, 4-hydroxy-3,5-dimethoxycinnamic acid) is a matrix used for proteins. Peptides <3000 Da may give weak signals (*see* **Note 11**).

11. If salts are present in the sample, then satellite peaks, containing Na^+ or K^+ that replace protons as the charge donor, may be seen. Similar satellite peaks will also occur when using poor quality matrix substances [20].
12. It is recommended to use 2:1 to 1:1 water–solvent ratios for matrix (*see* **Note 10**) solubilization.
13. SinA matrix should be stored in the dark in a desiccator. A ready to use SinA matrix solution should be wrapped in aluminum foil and used within 12 h.
14. Computer processing of a protein map allows to obtain a report. The report contains basic information about each separated protein spot (number of spot, p*I*, MW, intensity, area, volume of spot, % intensity, and % volume of spot). The results of such analysis can be compared with the results of a bioinformatic analysis.
15. The amino acid sequence is located in the "Sequence" line/category (*see* **Notes 16** and **17**). The sequence data correspond to the precursor form of a protein before posttranslational modifications and processing. Next to the written sequence there is a window in which the name of the first available program is placed. The database user, after opening the list of programs, can choose one out of six available programs (BLAST, ProtParam, ProtScale, Compute p*I*/MW, PeptideMass, PeptideCutter) (*see* **Note 18**).
16. The amino acid sequence is written in lines (50 amino acids each, in groups of 10). All the amino acids ending (at the C-terminal end of the sequence) each group of 10 amino acids are numbered.
17. The amino acid sequences are available in the following formats: a raw format, FASTA, and xml. Such record of a sequence enables a database user to copy the available sequence and use it for the next analysis, e.g., by using other computer programs.
18. The programs, i.e., ProtParam [2–4], Compute p*I*/MW [5–7], PeptideMass [37–39], are available on the website (http://web.expasy.org) [18]. With the use of the programs mentioned above one can analyze the amino acid sequence from the UniProtKB database (then you must give the AC or ID number) or another sequence. One should remember that the entered amino acid sequence must be written in a raw format (*see* **Notes 1, 19** and **20**). The amino acid sequence entered by the program user is automatically presented in lines. There are

60 amino acids in each line, in groups of 10. Above each (C-terminal) amino acid ending a group of 10 there is a number of subsequent tens (10, 20, 30, etc.).

19. It is possible to analyze one's own amino acid sequence written as a one-letter code (*see* **Notes 1** and **20**).
20. It is not possible to analyze the amino acid sequence with posttranslational modifications (e.g., after the addition of glycoside or lipid residues).
21. Theoretical MW and p*I* are calculated as in Compute p*I*/MW [5–7]. Protein MW is calculated by the addition of average isotopic masses of amino acids in the protein and the average isotopic mass of one water molecule. Molecular weight values are given in Dalton (Da). Protein p*I* is calculated using the p*K* values of amino acids described by Bjellqvist et al. [41, 42], which were defined by examining polypeptide migration between pH 4.5–7.3 in an immobilized pH gradient gel environment with 9.2 and 9.8 M urea at 15 or 25 °C. The prediction of protein p*I* for highly basic proteins is yet to be studied, and it is possible that current Compute p*I*/MW predictions may not be adequate for this purpose.
22. The buffer capacity of a protein will affect the accuracy of its predicted p*I*, with poor buffer capacity leading to a greater error in prediction [41, 42]. Because of this, p*I* predictions for small proteins can be problematic.
23. Hydrolysis of a protein/protein fragment (*see* **Notes 24–26**) can be performed by using available enzymes or combining the specificity of activity of two enzymes. The specificities of activity of individual enzymes are presented in the instruction [37–39].
24. Signal sequences, propeptides and transit sequences are all removed from the sequence of a protein or protein fragment before peptides are obtained.
25. All posttranslational modifications (e.g., acetylation, amidation, deamidation, formylation, hydroxylation, methylation, phosphorylation) which appear in the UniProtKB/Swiss-Prot database are taken into consideration by the program during the calculation of peptide masses.
26. In case of complicated modifications (including N- and O-glycosylation and phosphatidylinositol glycan anchors) it is difficult or impossible to calculate the mass of a modified peptide. Other restrictions are listed and described in the instructions for PeptideMass peptide characterization software [39].
27. This program is designed to calculate the theoretical masses of peptides generated by the chemical or enzymatic cleavage of proteins, to assist in the interpretation of peptide mass fingerprinting and peptide mapping experiments.

28. One can calculate the (average or monoisotopic) peptide masses whose amino acid residues have not been modified (in a non-reduced form) or in the protein there are modified cysteine residues (iodoacetic acid, iodoacetamide, 4-vinyl pyridene) or methionine residues (with methionines oxidized).
29. If you do not use a low-binding tube (Protein Lo-bind, Eppendorf or equivalent), prewash the tube twice with buffer A and let it dry completely.
30. Reduction and alkylation steps are skipped in this protocol because this was performed prior to the second dimension separation.
31. Tolerated levels of potential impurities: phosphate buffers <20 mM, tris(hydroksymethyl)aminomethane (tris) buffer <50 mM, detergents <0.1 %, alkali metal salts <1 M, guanidine <1 M, AmBi <30 mM. In the samples prepared for the MALDI-TOF analysis there should not be any glycerol, sodium dodecyl sulfate (SDS) (*see* **Note 32**), and sodium azide.
32. Strong ionic detergents, such as SDS, must be completely removed as they will interfere with crystallization, and so disrupt the mass analysis.
33. If polyacrylamide gels (carriers for the separation of proteins with the 2-DE technique) are prepared manually, they must be polymerized completely (it is best to prepare them a day before, before the separation of proteins in these polyacrylamide gels). This is of great significance especially if protein spots are cut out of them, and then samples for MS analysis are properly prepared (*see* **Note 34**).
34. A silver method of protein staining in a polyacrylamide gel according to Schevchenko [43] is one of staining methods that can be used if the MS analysis is performed next, but not with the use of MALDI-TOF MS. Glutaraldehyde (pentane-1,5-dial) or formaldehyde (methanal), i.e., compounds used in the silver method of protein staining, cannot be present in the samples prepared for analysis with the MALDI-TOF technique. Formaldehyde cross-links a protein in gel, which results in decreased efficiency of protein identification. The presence of this compound can also affect the adulteration of results obtained with the use of the MALDI-TOF MS technique [44].
35. When preparing a sample for analysis with the MALDI-TOF MS technique from a cut out protein spot obtained after the separation of proteins with 2-DE, one should also prepare a so-called control test. To this end, one should cut out a fragment of polyacrylamide gel (from a place where there are no separated protein spots) corresponding to the size of the analyzed protein spot, which makes it possible to eliminate potential artifact peaks.

36. Rinse the scalpel well after excising each spot.
37. Protein spots on polyacrylamide gels (after the separation of proteins with the 2-DE technique) can be cut out manually (with a scalpel), or automatically with a spot picker for cutting out protein spots. This device can be equipped with a system that makes it possible to perform protein digestion (the stage preceding the placing of a sample on the MALDI matrix). In case of protein digestion performed directly in gel, the samples are immediately transferred on the matrix [45, 46].
38. In proteomic research the most often used MS technique is the combination of ion sources MALDI and the analyzer of the time of flight (TOF). Ions are analyzed with the TOF detector, where heavier ions reach the detector later than lighter ions. At the end of the distance covered by the ions in the TOF detector, ions with the same m/z values are focused with a system of electronic lenses (reflectron), and turned back to the detector. The reflectron focuses the ions, improving the resolution of analyses, and the distance covered by ions is nearly doubled, which additionally increases the resolution of measurements. The result of analysis is always the m/z value [19, 20].
39. The quality of obtained results primarily depends on the purity of the analyzed sample, measurement accuracy and the number of measured peptides. A protein from a single spot isolated from polyacrylamide gel in the amount <1 ng can be identified.
40. Mascot is the best-known tool-program for protein identification [30, 31]. Users can select a database which the analyzed amino acid sequences of proteins come from (*see* **Note 41**). They can also select the enzyme used in the experiment, the species of identified proteins, modifications of proteins. They obtain a list of proteins listed according to the results based on the scoring function. In addition, the significance level for the highest result is given.
41. In order for a protein to be identified, its sequence or sequences of similar proteins must be present in the databases. If a protein sequence is not deposited in the databases, programs used for protein identification (on the basis of a peptide map obtained thanks to the MALDI-TOF technique) will not indicate the right protein, despite obtaining reliable mass spectra.
42. During the matching of theoretical masses to experimental masses one should remember about the so-called doubling of peptide masses. Such amino acids as leucine and isoleucine have identical molecular masses, which means that two peptides with the same amino acid sequence, which differ only in these amino acids, will have the same mass. Moreover, as a result of in silico hydrolysis of some sequences of proteins which come from protein databases, also peptides with identical masses can be the

products of reactions. In case of obtaining an experimental spectrum of a peptide map it is not possible to tell from which amino acid sequence the peptide mass comes from.

43. One should remember that there are many factors affecting the masses of peptides observed in a mass spectrum. One group of these factors consists of impurities which can appear at the stage of sample preparation (*see* **Notes 3–5**, **31** and **32**). The other group consists of protein modifications (*see* **Note 44**). All the factors affecting the masses of obtained peptides should be taken into consideration during the calculation of the masses of peptides obtained from the analyzed amino acid sequence of a protein (which comes from protein databases). As a result of hydrolysis of peptide bonds in proteins by proteinases, in the calculation of the masses of obtained peptides one should take into consideration the masses of hydrogens (H) and hydroxide groups (OH) located at N-terminals and C-terminals of peptides respectively [47].
44. One should remember that all modifications of proteins performed at the stage of preparation of a sample for analysis with a mass spectrometer or at the stage of 2-DE separation of proteins (a stage used before the MS analysis) should be considered by calculating the theoretical masses of peptides obtained as a result of in silico hydrolysis of proteins. Such modifications include the modification of cysteine residues under the influence of applied iodoacetamide. The modification of cysteine takes place also when the sample contains acrylamide, which is a component of polyacrylamide gel used during electrophoresis (*see* **Note 33**). After the addition of acrylamide to cysteine residues, a modification of the mass of this amino acid takes place. Other amino acid modifications include for example acetylation of the N-terminal or lysine, phosphorylation (serine, threonine, tryptophan), formation of disulfide bridges, oxidation of histidine, methionine, tryptophan.
45. Take care during sample handling: protein adsorption to surfaces such as pipettes or Eppendorf tubes can significantly reduce the true sample concentration, particularly if the original concentration is low.

Acknowledgment

This work was supported by Ministry of Science and Higher Education, Poland within project No NN 312 286366.

References

1. The Universal Protein Resource (UniProt) (2002). http://www.uniprot.org. Accessed Nov 2015
2. The ProtParam program. http://web.expasy.org/cgi-bin/protparam/protparam. Accessed Nov 2015
3. The ProtParam program. http://web.expasy.org/protparam/. Accessed Nov 2015
4. The ProtParam. Documentation is available at http://web.expasy.org/protparam/protparam-doc.html. Accessed Nov 2015
5. The Compute *pI/MW* program. http://web.expasy.org/cgi-bin/compute_pi/pi_tool. Accessed Nov 2015
6. The Compute *pI/MW* program. http://web.expasy.org/compute_pi/. Accessed Nov 2015
7. The Compute *pI/MW*. Documentation is available at http://web.expasy.org/compute_pi/pi_tool-doc.html. Accessed Nov 2015
8. The UniProt Knowledgebase user manual. http://web.expasy.org/docs/userman.html. Accessed 20 Nov 2015
9. Pundir S, Magrane M, Martin MJ et al (2015) Searching and navigating UniProt databases. Curr Protoc Bioinform 50:1.27.1–1.27.10. doi:10.1002/0471250953.bi0127s50
10. Magrane M, UniProt Consortium (2011) UniProt Knowledgebase: a hub of integrated protein data. Database 2011:bar009. doi:10.1093/database/bar009
11. Jain E, Bairoch A, Duvaud S et al (2009) Infrastructure for the life sciences: design and implementation of the UniProt website. BMC Bioinformatics 10:136. doi:10.1186/1471-2105-10-136
12. The UniProt Consortium (2008) The Universal Protein Resource (UniProt). Nucleic Acids Res 36(Database issue):D190–D195. doi:10.1093/nar/gkm895
13. Boutet E, Lieberherr D, Tognolli M et al (2007) UniProtKB/Swiss-Prot. The manually annotated section of the UniProt Knowledgebase. In: Edwards D (ed) Methods in molecular biology, plant bioinformatics: methods and protocols. Humana Press Inc, Totowa, NJ, pp 89–112
14. Boeckmann B, Blatter MC, Famiglietti L et al (2005) Protein variety and functional diversity: Swiss-Prot annotation in its biological context. C R Biol 328(10–11):882–899. doi:10.106/j.crvi.2005.06.001
15. Apweiler R, Bairoch A, Wu CH (2004) Protein sequence databases. Curr Opin Chem Biol 8(1):76–80. doi:10.1016/j.cbpa.2003.12.004
16. Apweiler R, Bairoch A, Wu CH et al (2004) UniProt: the Universal Protein knowledgebase. Nucleic Acids Res 32(Database issue):D115–D119. doi:10.1093/nar/gkh131
17. Bairoch A, Boeckmann B, Ferro S et al (2004) Swiss-Prot: juggling between evolution and stability. Brief Bioinform 5(1):39–55. doi:10.1093/bib/5.1.39
18. The Swiss Institute of Bioinformatics (SIB) ExPASy (Expert Protein Analysis System). Bioinformatics Resource Portal. http://www.expasy.org. Accessed Nov 2015
19. Amersham Biosciences (2003) Ettan MALDI-TOF Pro version 2.0 user manual. Uppsala, Sweden
20. Amersham Biosciences (2003) Ettan MALDI-TOF method handbook. Uppsala, Sweden
21. Tanaka K (2003) The origin of macromolecule ionization by laser irradiation (Nobel lecture). Angew Chem Int Ed 42(33):3860–3870. doi:10.1002/anie.200300585
22. Forner F, Foster LJ, Toppo S (2007) Mass spectrometry data analysis in the proteomics era. Curr Bioinformatics 2(1):63–93. doi:10.2174/157489307779314285
23. Trauger SA, Webb W, Siuzdak G (2002) Peptide and protein analysis with mass spectrometry. Spectroscopy 16(1):15–28. doi:10.1155/2002/320152
24. Hickman DR, Roepstorff P, Shewry PR et al (1995) Molecular weights of high molecular weight (HMW) subunits of glutenin determined by mass spectrometry. J Cereal Sci 22(2):99–103. doi:10.1016/0733-5210(95)90038-1
25. Lim H, Eng J, Yates JR et al (2003) Identification of 2D-gel proteins: a comparison of MALDI/TOF peptide mass mapping to mu LC-ESI tandem mass spectrometry. J Am Soc Mass Spectrom 14(9):957–970. doi:10.1016/s1044-0305(03)00144-2
26. Salzano AM, Crescenzi M (2005) Mass spectrometry for protein identification and the study of post translational modifications. Ann Ist Super Sanita 41(4):443–450
27. Camafeita E, Solís J, Alfonso P et al (1998) Selective identification by matrix-assisted laser desorption/ionization time-of-flight mass spectrometry of different types of gluten in foods made with cereal mixtures. J Chromatogr A 823(1–2):299–306. doi:10.1016/s0021-9673(98)00621-9
28. Henzel WJ, Watanabe C, Stults JT (2003) Protein identification: the origins of peptide mass fingerprinting. J Am Soc Mass Spectrom 14(9):931–942. doi:10.1016/s1044-0305(03)00214-9
29. Thiede B, Höhenwarter W, Krah A et al (2005) Peptide mass fingerprinting. Methods

35(3):237–247. doi:10.1016/j.ymeth.2004.08.015

30. Perkins DN, Pappin DJ, Creasy DM et al (1999) Probability-based protein identification by searching sequence databases using mass spectrometry data. Electrophoresis 20(18):3551–3567. doi:10.1002/(sici)1522-2683(19991201)20:18<3551::aid-elps3551>3.0.c.o;2-2
31. Matrix Science. http://www.matrixscience.com/search_form_select.html
32. Galvani M, Hamdan MAA, Righetti PG (2001) Two-dimensional gel electrophoresis/matrix-assisted laser desorption/ionization mass spectrometry of commercial bovine milk. Rapid Commun Mass Spectrom 15(4):258–264. doi:10.1002/rcm.220
33. Lopez-Ferrer D, Canas B, Vázquez J et al (2006) Sample treatment for protein identification by mass spectrometry-based techniques. Trends Anal Chem 25(10):996–1005. doi:10.1016/j.trac.2006.05.015
34. Promega in-gel protein digestion protocol. Trypsin gold, mass spectrometry grade. Technical bulletin available at http://www.promega.com/~/media/Files/Resources/Protocols. Accessed Nov 2015
35. National Center for Biotechnology Information (NCBI) database. http://www.ncbi.nlm.nih.gov/protein
36. Gasteiger E, Hoogland C, Gattiker A et al (2005) Protein identification and analysis tools on the ExPASy server. In: Walker JM (ed) The proteomics protocols handbook. Humana Press, Totowa, NJ, pp 571–607. doi:10.1385/1-59259-890-0:571
37. The PeptideMass program. http://web.expasy.org/cgi-bin/peptide_mass/peptide-mass.pl. Accessed Nov 2015
38. The PeptideMass program. http://web.expasy.org/peptide_mass/. Accessed Nov 2015
39. The PeptideMass. Instructions for PeptideMass peptide characterization software. http://web.expasy.org/peptide_mass/peptide-mass-doc.html. Accessed Nov 2015
40. Levander F, Rögnvaldsson T, Samuelsson J et al (2004) Automated methods for improved protein identification by peptide mass fingerprinting. Proteomics 4(9):2594–2601. doi:10.1002/pmic.200300804
41. Bjellqvist B, Hughes GJ, Pasquali C et al (1993) The focusing positions of polypeptides in immobilized pH gradients can be predicted from their amino acid sequences. Electrophoresis 14(1):1023–1031. doi:10.1002/elps.11501401163
42. Bjellqvist B, Basse B, Olsen E et al (1994) Reference points for comparisons of two-dimensional maps of proteins from different human cell types defined in a pH scale where isoelectric points correlate with polypeptide compositions. Electrophoresis 15(1):529–539. doi:10.1002/elps.1150150171
43. Schevchenko A, Wilm M, Vorm O et al (1996) Mass spectrometric sequencing of proteins silver-stained polyacrylamide gels. Anal Chem 68(5):850–858. doi:10.1021/ac950914h
44. Scheler C, Lamer S, Pan Z et al (1998) Peptide mass fingerprint sequence coverage from differently stained proteins on two-dimensional electrophoresis patterns by matrix assisted laser desorption/ionization-mass spectrometry (MALDI-MS). Electrophoresis 19(6):918–927. doi:10.1002/elps.1150190607
45. Canelle L, Pionneau A, Marie A et al (2004) Automating proteome analysis: improvements in throughput, quality and accuracy of protein identification by peptide mass fingerprinting. Rapid Commun Mass Spectrom 18(23):2785–2794. doi:10.1002/rcm.1693
46. Katayama H, Nagasu T, Oda Y (2001) Improvement of in-gel digestion protocol for peptide mass fingerprinting by matrix-assisted laser desorption/ionization time-of-flight mass spectrometry. Rapid Commun Mass Spectrom 15(16):1416–1421. doi:10.1002/rcm.379
47. Song Z, Chen L, Ganapathy A et al (2007) Development and assessment of scoring functions for protein identification using PMF data. Electrophoresis 28(5):864–870. doi:10.1002/elps.200600305

Index

Sebastian Gasparis (ed.), *Oat: Methods and Protocols*, Methods in Molecular Biology, vol. 1536, DOI 10.1007/978-1-4939-6682-0, © Springer Science+Business Media LLC 2017

MIX
Papier aus verantwortungsvollen Quellen
Paper from responsible sources
FSC® C105338

If you have any concerns about our products,
you can contact us on
ProductSafety@springernature.com

In case Publisher is established outside the EU,
the EU authorized representative is:
Springer Nature Customer Service Center GmbH
Europaplatz 3, 69115 Heidelberg, Germany

Printed by Libri Plureos GmbH
in Hamburg, Germany